AF836687

ORNAMENTAL PLANTS

NIPA® GENX ELECTRONIC RESOURCES & SOLUTIONS P. LTD.
New Delhi-110 034

www.nipaers.com
Browse, Search, Read & Buy...
Print Books
eChapters
eBooks
Publishers
Forthcoming
Subject Catalogues
New Titles
Online Resources on:
✓ Current Affairs
✓ Reasoning, Logic & Aptitude
✓ Competitive Examinations
✓ English Language Lab
✓ Writing & Pronunciation Tools
✓ Personality Development
✓ Online Programmes for Professionals
✓ Interview Preparation
Pay Using
PhonePe
Google Pay
paytm
PayPal
UPI
CCAvenue

ORNAMENTAL PLANTS

Authored by

Sabina George Thekkayam

Professor

Department of Pomology and Floriculture

Kerala Agricultural University, Thiruvananthapuram

Kerala, India

Edited by

Prof. K.V. Peter

Director

World Noni Research Foundation

Chennai, Tamil Nadu, India

NIPA® GENX ELECTRONIC RESOURCES & SOLUTIONS P. LTD.

New Delhi-110 034

NIPA® GENX ELECTRONIC
RESOURCES & SOLUTIONS P. LTD.

101,103, Vikas Surya Plaza, CU Block
L.S.C. Market, Pitam Pura, New Delhi-110 034
Ph : +91 11 27341616, 27341717, 27341718
E-mail: newindiapublishingagency@gmail.com
web: www.nipabooks.com

For customer assistance, please contact
Phone: + 91-11-27 34 17 17
Fax: + 91-11- 27 34 16 16
E-Mail: feedbacks@nipabooks.com

ISBN: 978-81-19235-98-8

NIPA® also publishes books in a variety of electronic formats. Some content that appears in print may not be available in electronic books, and vice versa.

Composed and Designed by NIPA®.

Dedicated to

My Husband

DR. M. L. JOSE

&

Daughters

PREETHI & MINU

FOREWORD

Plant lovers are constantly on the lookout for newer plants for their gardens. With its wide variety of geoclimatic conditions the tropical indian gardens provide a variety of habitats for raising local and introduced plants. Our gardens present floral elements of the Himalayas, the northeastern belt, the river deltas, the montane and sub-montane tracts of north and central India, the southern plateau, the littoral regions and the western and eastern ghats. Many ornamentals previously confined to particular geographic areas are now distributed ubiquitously and have been brought into our gardens and landscapes. Oving to the recent advances in growing and irrigation techniques of ornamental plants surmounting evolutionary and reproductive barriers, a bewildering number of new plants are being produced every year. Many of these exist only in gardens or high-tech commercial growing centers.

Human ingenuity created diversity of gardens offers a plant mix predominantly ornamental. From the array of plants that are useful to us, garden plants and flowers are the most familiarized with. The benefits derived from growing plants and flowers in gardens are unique. In addition to personal adornment or worship, globalization has brought in interior floral decoration, social offerings and for decorative purposes into a business proposition. They touch everyday life, enlightening us and enlivening our existence. As living symbols of growth, development, senescence, birth and growth of fruits, seeds, regeneration and survival-all these are symbolic of nature's processes. They are also silent, self sustaining and aesthetically pleasing living beings we co-exist with. A mere stroll in the garden makes our heart young and makes us a part of nature.

All the species in the plant kingdom were wild. The group called 'ornamentals' is a minority. Their shapes, colouration, markings on the petals and scent were meant for attracting pollinating insects like bees, butterflies and birds. They are rewarded with pollen and nectar for the pollination services. Humans take advantages of these phenomina. They selected, domesticated and bred hundreds of varieties of ornamentals, many of which lost in the process - the natural environment to set seeds. A large number of garden plants are propagated by stem cuttings, root suckers, bulbs, rhizomes, corms and even leaves. But they sustain the economically viable sector of ornamental horticulture.

Dr. Sabina George has recorded the popular ornamental plants of Indian gardens in a commendable way.

I appreciate Dr. K.V. Peter, Former Vice-Chancellor, Kerala Agricultural University for encouraging the author to take up this work and for succeeding to put into print another valuable book on Ornamental Horticulture. The effort is commendable and it is my hope this book will help users to identify garden plants and select plants for building up gardens to bring joy and transquility.

Sd/-

Dr. H.Y. Mohan Ram
INSA Honorary Scientist
Formerly Professor of Botany
University of Delhi, Delhi

PREFACE

India is home to a large number of plants that are grown ornamentally throughout the world. Climatic and geographic features have moulded the subcontinent's floristic wealth. The germplasm of ornamental plants available in India is phenomenally extensive and probably as diverse. The introduction of plant species that has taken place over centuries is the main causative factor for this. Throughout the world, the fascination for novelty has led to the spread of ornamental plants into places far removed from their centers of origin. Such distribution has far exceeded natural migration and dispersal of plants. Though this inhabitation and acclimatization in newer areas similar to their natural habitats can be considered successful from the point of view of distribution, many plants have been led to the verge of extinction in their original homes due to over collection for commercial dispersal. Deforestation and unscientific afforestation have endangered many plants especially the epiphytic, lithophytic and shade adapted native species of precariously balanced survival niches all over the world. This has necessitated efforts towards conservation of native habitats of unique endemic species in biodiversity "hot spots". A recent example of this is the declaration of the "Kurinjimala Sanctuary" as the protected bio-reserve of Neela Kurinji (*Phlebophyllum kunthianum* Nees). The ecological impact of uncontrolled plant introductions on the populations of native species is considerable as they invariably replace them, affecting the biological balance in the ecosystem. Many a time, it is only when they attain the status of weeds that their impact is noticed. Ornamentals are found to have encroached unnoticed into our evergreen forests too. The African Tulip tree (*Spathodea campanulata*) has been introduced reliberately by planters to the Western Ghats of Kerala and can be seen dotting the blue-green hills with their flame-red flowers during September-October. A fairly recent migrant that has a record of introduction into neighbouring Sri Lanka is the Jamaican Cherry (*Muntingia calabura*). This small tree with its low horizontally oriented branches that spread like canopy is presently widely distributed from the coastal to the midland regions of Kerala and is also found growing along the roadsides fringing the Ghats. The fruits are eaten and the seeds dispersed by birds.

The varying cycles of acceptability enjoyed by ornamentals is a cause for the steady influx of new plants into new habitats as well as their erosion from them. From researchers to amateur gardeners, people associated with ornamental plant growing are aware of the speed with which a plant becomes 'popular' or falls from favour due to becoming 'common'.

In this context, it was felt that the vastness of ornamental plants germplasm presently available in the country needed horticultural validation. This book is thus an attempt to put on record majority of the ornamental plants grown throughout India. The descriptions given are a record of the morphological characteristics of the plants as observed in growing situations in gardens. To a student of ornamental gardening and landscaping this work is intended to be a practical guide in identifying and distinguishing the ornamental features of the enumerated plants. To amateur and professional gardeners and landscape architects this is intended to be a handbook for selection of ornamentals for specific functions and locations. This work may also serve as an eye opener to conservationists and plant scientists who are concerned about the dwindling native flora and the increasing number of introduced plants in Indian gardens and designed landscapes.

Thiruvananthapuram
30-04-2009

Dr. Sabina George T.
K.V. Peter

ACKNOWLEDGEMENT

I wish to record my profound gratitude to Dr. K. V. Peter, former Vice-Chancellor, Kerala Agricultural University for being a patient and untiring motivator of this book and for bearing with me throughout the turbulent period of it's writing. Dr. H.Y. Mohan Ram, INSA Honorary Scientist and Formerly Professor of Botany, University of Delhi for the scholarly Foreword to the book.

I am indebted to my colleagues who have inspired me by their enthusiasm about ornamentals and gardens and to my students who have nurtured my search for plants and names by their persistent queries and inquisitiveness about the native flora of our country. This arduous academic work would not have materialized but for the support and encouragement extended by my husband Dr. M.L. Jose and daughters Preethi and Minu. I devote this book for their patience and forbearance.

The photographs are of the ornamental plants maintained in the gardens of the College of Agriculture, Vellayani, Thiruvananthapuram.

The library and literature search facilities provided by the Kerala Agricultural University are gratefully acknowledged.

Thiruvananthapuram
30-04-2009

Dr. Sabina George T.

ABBREVIATIONS

Language Name	Abbreviation
Assamee	Assa.
Bengali	Beng.
English	Eng.
Gujarati	Guj.
Malayalam	Mal.
Marathi	Mar.
Punjabi	Punj.
Sanskrit	Sans.
Tamil	Tam.
Telugu	Tel.

INTRODUCTION

The plant wealth of Indian gardens conforms to our country's rich heritage and exemplifies the benefits of cultures that have enriched ours through contributions of ornamental flora. The acclimatization, propagation and popularization of indigenous and introduced garden plants throughout the country has been greatly aided by their being grown in the relatively well cared for environments of gardens. However in the never-ending quest for newer plants, many of the indigenous species have been replaced by newer introductions. Very few of the ornamentals that graced the gardens of the last millennium are found in today's gardens. Many such "relics" survive in Botanic Gardens and in very large public parks only.

Diversity of Plants in Gardens

Plants are the most ideal living species for co-habitation in human environments. From their natural habitats outdoors, they have been brought indoors into living areas by human beings to derive their aesthetic and other beneficial effects. In gardens and structured landscapes, plants of different kinds are grown for display by forming lawns, avenues, beds, borders, edgings etc. They are also trained and pruned to form hedges, standards, arches, pergolas and various kinds of topiary. In the tropics, plants suited to various tropical ecosystems can be successfully grown for ornamentation by recreating ecological niches to which they are adapted. A rockery accommodating hot-desert plants and a fernery of tropical ferns, selaginellas and mosses can easily find place in a tropical garden. In the humid tropics, an Alpine garden containing Araucarias, Junipers, Thujas and even the Christmas holly can be laid out. The wide adaptability of plants can thus be skillfully used to exhibit their diversity and aesthetic value in gardens.

Display Effects of Garden Plants

From tall growing trees to ground hugging creepers and trailers, a variety of plants are grown in gardens and landscapes. Annual and perennial flowering herbs as well as herbaceous perennials and bulbous plants are ideal for group planting in order to obtain 'mass effects'. Small leaved foliage shrubs and flowering shrubs make excellent hedges while arboreal shrubs and small trees can be trained as screens and barriers. Mixed plantings of plants of more or less similar stature give an effect of a natural stand of plants like '**Tree groves**' and ' **Mixed borders**'. '**Ground-cover**' plants are also used to mask the barrenness of open surfaces such as the ground, walls, rocks and even bare tree-trunks. Plants are also made even more appealing by topiary.

Plant Groups of Gardens

Plants grown in gardens and landscapes can be grouped variously. As they are primarily entities having differing dimensions and occupy space differently, they can be grouped based on their habit as **Trees**, **Shrubs**, annual and perennial **Herbs**, **Climbers** and **Creepers**. In these groups, several plant families of dicots and monocots are included.

In addition to these, certain plants are grown collectively in gardens due to their similarity in structural features. Such groupings may consist of plants, which differ greatly in height, stem girth, canopy structure and the shape, size and orientation of branches and in floral characters. An example is a grouping that contains **Cycads, Tree Ferns** and **Palms**. However most of the groupings consist of related plants, which share morphological and physiological features. Examples are '**Cacti** and **Succulents**', '**Grasses'**, '**Bamboos' 'Sedges** and **Rushes**', '**Bulbs**', '**Orchids' 'Bromeliads'**, '**Ferns** and **Selaginellas**'. An enhanced visual effect from their grouping and the impression of a natural stand in addition to easiness in management and assessment of comparative performance are the beneficial effects of such groupings. Another broad group are the '**Ornamental Foliage Plants'.** These are grown primarily for their attractive foliage. They are extensively used in tropical gardens and indoor areas and are spectacularly displayed as potted plants. Many are made into hedges and are used as ground covers and foundation covers and for making topiary.

Different kinds of plants including trees, shrubs and succulent herbs are selected for growing as ornamental foliage plants. Characteristics such as leaf colour and variegations, leaf shape, structure and orientation are used for selection. Foliage plants include monocots and dicots having woody stems like trees and shrubs, and non-woody or succulent-stemmed plants like perennial herbs, creepers and climbers and also those possessing modified stems like the bulbous plants. Woody monocots with tree-like habit like the palms and bamboos and the more primitive cycads, ferns and selaginellas as well as the grasses, cacti and succulents that are primarily grown for the decorative value of their leaves and plant form are as mentioned above, dealt with separately under respective heads. The popularly grown foliage plants of the tropics can thus be categorized based on their habit into Monocotyledonous Foliage Shrubs, Dicotyledonous Foliage Shrubs, Foliage Herbs and Foliage Climbers and Creepers.

Another group that is grown separately due to their specific garden habitats are the **Water Garden plants**. Similar habitat-specific groups also include the epiphytic ferns and orchids, and mosses and other lithophytes. Ferneries, Orchidaria and Moss Gardens are also unique specialty gardens of today.

CONTENTS

Chapter 1

CARE OF TREES

Trees are the largest plant components of gardens and landscapes. However much architectural features dwarf them, for human beings, they are the largest living entities in the environment. Profoundly influencing our habits, styles of habitation, culture and daily life, trees have been revered, sanctified and given symbolic importance and above all, depended upon as means of livelihood by entire communities through the ages. Through folklore, traditional knowledge and rituals, the usefulness of trees has been handed down from generation to generation-in societies all over the world. In India, cave paintings and carvings in addition to ancient literary works document the aesthetic and medicinal knowledge of our forefathers about trees found in the sub-continent.

Trees differ in stature, branching habit, girth of trunk and branches, canopy structure, outline and crown shape. The colour of the foliage, its texture, glaze, shape and size, foliage colour, floral features and season of flowering also differ among trees. These characteristics in addition to giving distinctiveness, also determine their ornamental use in gardens and landscapes. In tropical gardens evergreen dicotyledons and palms dominate among the trees, whereas in temperate situations evergreen conifers are selected over the deciduous dicots for tree planting. Palms among trees are universally grown as avenue trees or as clumps in areas where they can be easily grown while the other trees are grown as such or pruned and trained into various forms or as bonsai in gardens.

CLASSIFICATION OF TREES

1. Classification Based on Stature

The stature attained by trees at maturity enables them to be classified as Large trees, Medium-large trees and Small trees. Large trees are those, which attain heights of over 15m. Medium large are those, which attain heights of 8 to 15m and Small trees are those that attain less than 8m height.

Large trees : *Azadirachta indica, Ficus religiosa, Ficus bengalensis, Delonix regia, Enterolobium saman, Couropita guianensis, Casuarina equisetifolia, Millingtonia hortensis, Grevillea robusta.*

Medium large trees : *Nyctanthes arbor-tristis, Jacaranda mimosaefolia, Lagestroemia flos-reginae, Filicium decipiens, Ficus retusa, Crataeva nurvala.*

Citharexylum quadrangulare, Cassia fistula.

Small trees : *Muntingia calabura, Saraca indica, Melia azaderach, Caesalpinea coriaria, Pisonia alba.*

2. Classification Based on Canopy Outline and Form

The arrangement of branches and the size, shape, texture and orientation of leaves give trees a particular form and canopy outline. Trees that attain their maximum stature unencumbered, are able to develop their distinctive characteristics fully. Many coniferous trees exhibit an architectural form, which enable them to be classed as having pyramidal or conical (Araucarias, Cypresses), fastigiated, or columnar (*Polyalthia longifolia* var. *pendula,* Junipers) outlines. In most of the dicotyledonous trees, canopies are spreading (*Samanea, Delonix*, Divi-divi), open-headed (*Ficus religiosa, Jaccaranda,* the Cassias, *Tabebuia*), or rounded or dome-shaped (Palms, Cannon-ball tree, the Fiddle-wood tree). A weeping crown with drooping branches and pendant leaves are unique to contain species of pines, willows and cypresses and to the Bottlebrush and the Weeping fig.

3. Classification Based on the Decorative Nature of Leaves and Flowers

Trees with inconspicuous flowers but having ornamental foliage are called **Foliage trees**. They are primarily grown for shade and the decorative nature of their foliage cover.

Trees with conspicuously ornamental flowers are called **Flowering trees**. Their nature of flowering and the size, colour and number of blooms per group or bunch make them spectacular or be noticed for their flowers.

4. Classification Based on Floral Characteristics

Though trees are large entities, the flowers of very few trees are more than 3cm in diameter. Relatively large sized flowers are seen in the Tulip tree, the Portia tree, the Cannon Ball tree, the Gul Mohr and in the jacarandas. Most trees have small flowers in fairly small sized clusters produced all over the canopy. The size of the clusters and their colour contrasts with the foliage or the absence of the foliage at flowering make them conspicuous and more or less attractive. Spectacular flowering trees of the tropics, which flower in a decidous or semi-decidous state, include the cassias, tecomas, the jacarandas and the sacred Barna.

a) Classification based on the season of flowering

The months of March to May which heralds Spring in Northern India and in the hills of Central and Southern India and at the same time denotes the onset of summer in the rest of India is the main flowering season of trees in the sub-continent. The cassias, *Bauhinia, Lagestroemia, Michelia*, Gul mohr, Asokam, and *Citharexylum* are among the spring/summer flowering trees. In response to the lowering of temperatures in December, flowering takes place by January in *Amherstia, Bombax* and *Mimusops*. The rainy season from June to August sets off flowering in *Couropita* and *Nyctanthes* and a second burst of flowers in *Peltophorum* and *Michelia*.

b) Classification based on floral scent

Intoxicating fragrances to pungent aromas are associated with flowers of trees. They may be delicately scented as in the honey-scented flowers of Neem and Persian lilac or lightly scented as in the flowers of cassias. Strongly scented are the flowers of the Cannon - Ball tree, the Golden Champa, the Coral Jasmine, the Ylang-ylang and the Plumerias. Among fragrant flowered trees unique with a spicy scent are the flowers of *Alstonia scholaris*.

5. Classification Based on Foliage Characteristics

Based on foliage size, surface texture and leaf orientation, trees may have densely filled crowns as in the Copper-pod tree, the Noble Amherstia and the

Cannon - Ball tree while the crown may be sparsely filled as in the Bottle brush or the *Casuarina*. The surface texture presented by the foliage may be coarse as in the Cannon-Ball tree, the Cassias and the *Citharexylum* or fine textured as in *Casuarina*, the Divi-divi, the Junipers and the Thujas.

6. Classification Based on Suitability for Gardening and Landscaping

a) For roadsides and highway avenues

Large flowering trees: *Peltophorum*, Pink Cassias, Indian Laburnum, *Lagestroemia*, *Spathodea*.

Small flowering trees : *Bauhinia, Michelia, Tabebuia, Saraca*

Large foliage trees: *Enterolobium* , *Azadirachta, Adenanthera, Grevillea, Mimusops*, Tamarind

Small foliage trees: *Muntingia*, Divi-divi, *Pisonia alba* and small leaved Ficuses.

b) For central avenue planting

Polyalthia longifolia var. *pendula*, Traveller's palm, *Araucaria* spp. *Cupressus* spp.

c) Group planting for Grove-effect

Cycads, suckering Palms, Pines.

d) As specimen plants

The Baobab tree, the Bodhi trees, the Cannon-ball tree, the Pagoda trees.

e) For Bonsai

Cupressus spp., *Ficus* spp., *Pinus* spp.

FLOWERING AND FOLIAGE TREES OF INDIAN GARDENS

Ornamental flowering trees and foliage trees popularly grown in gardens and used for landscaping in India are listed below:

Adansonia digitata L. Bombacaceae

African Calabash tree, Baobab tree, Monkey bread tree (Eng.), Gorak Amli, Gorak Lichora (Hindi/Guj.), Gorak Chinch (Mar.), Anaipuli (Tam.).

Trees are deciduous and native to the western coast of Africa. They are considered to be very long living. The trunk attains disproportionate stoutness- of up to 10m - and tapers upwards abruptly. The bark is smooth and gray. Leaves are palmate. The large white flowers are about 15 cm in diameter and have purple stamens.

Fruits resemble long gourds and are eaten by monkeys. Trees were introduced into Botanic gardens in India as specimen plants. But in cultivation the trunk does not attain the spectacular proportions observed in its native habitat.

Adenanthera pavonina L. Fabaceae

Bead tree, Coral-wood tree (Eng.), Rakhta kambal (Beng.), Moti Chanothi (Guj.), Ratan Gunj, Thorli Gunj (Mar.), Bandiguruvenda (Tel.),Anilkundumani (Tam.), Manjadi (Mal.).

Trees have spreading branches and have a rough grayish coloured bark. They are native to China and Malaya. The heartwood is coral red in colour. The leaves are dark green and bipinnate. The yellow fragrant flowers are produced in axillary panicles from March to May. The long curved pods contain scarlet lenticular seeds. The pods dehisce on the tree and the two halves contort and curl exposing the bright red seeds adhering to the silvery white inner surface of the pods. These are popular avenue trees of public parks and gardens.

Alstonia scholaris (L.) R. Br. Apocynaceae

Devils tree, Dita bark (Eng.), Saptaparni (Sans.), Chatian (Hindi), Chattim (Beng), Satvin (Mar.), Palaigh (Tel., Tam.), Eyzhilam Paala (Mal.).

A tall elegant tree attaining a height of up to 15 m with horizontal branches in whorls making it an ideal shade tree. The basal portion of the trunk becomes buttressed with age. The flowers are villous, greenish white opening at night and have a pungent spicy scent. Leaves are usually seven in a whorl. All plant parts contain bitter alkaloids. Not preferred for planting in gardens due to its legendary association with evil spirits. Hence, confined to roadsides and untended lands in the humid tropical belt of South India.

Amherstia nobilis Wall. Fabaceae

Noble Amherstia, Queen of Flowering trees

Trees are evergreen and attain heights of up to 15m and carry a dense crown of elongated pinnate leaves which when young are coppery pink and contrast beautifully with the mature dark green leaves. The orange-red flowers are tipped yellow and hang in pendant racemes. These flaming and drooping inflorescences make *Amherstia* one of the most beautiful trees of gardens. Though plants thrive in the high rainfall belt of South India, the fail to flower as profusely as in the North.

Amoora rohituka (Roxb.) Wight & Arn. Meliaceae

Sohaga (Hindi), Amoor (Beng.).

A tall evergreen adapted to moist shady locations along the western coast of peninsular India. The leaves are oblong, lanceolate, imparipinnate and in 4 to 8 pairs along with the odd leaflet. The flowers are small, white, borne on stiff axillary spikes. The fruits are globular or egg shaped. They split open when ripe, exposing the deep orange coloured aril.

Anthocephalus cadamba Miq. Rubiaceae

The Kadamba tree (Eng.), Kadamba (Hindi / Guj.), Kadambamu (Tel.), Attutek (Mal.).

A large deciduous tree attaining a height of up to 20m with a spreading crown and drooping branches. The leaves are long stalked and cordate-ovate in shape. The globose yellow flower heads are delicately scented and emerge in June-July. They are considered to be harbingers of the monsoons. The fruits are creamy yellow and globose, shaped like an orange. In India, trees are revered due to their association with Lord Krishna. They are highly susceptible to frost.

Araucaria cookii R. Br. ex D. Don Araucariaceae
syn. *Araucaria columnaris*

Araucaria, Cook Pine

An elegant conifer with horizontally oriented symmetric branches carrying dark green closely spaced claw like needles. Plants thrive in the tropical belt of South India and attain a height of up to 60m but produce cones at high altitudes only. The architectural features of this tree make it a prized plant of home gardens. In India, trees are majestically displayed in avenues and as specimen plants in public parks and gardens. Other species introduced into India having similar canopy structure include the Bunya-bunya Pine *Araucaria bidwillii* Hook. and the Norfolk Island Pine *Araucaria heterophylla*

Azadirachta indica A. Juss. Meliaceae

Margosa (Eng.), Neem (Hindi / Beg.), Limdo (Guj.), Arya Veppu (Tel./ Tam./ Mal.).

Trees are the best-known among the medicinal trees of India. They are considered to be sacred and are grown throughout the country. They are medium to large sized attaining heights of up to 15 m. Leaves are imparipinnate, lanceolate and oblique with a broadly toothed margin. White honey scented flowers are produced in small axillary panicles. The oval fruits are single seeded. They are relished by birds when ripe. All plant parts except the ripe pericarp is bitter in taste. Trees are ideal for avenue planting along highways.

Bauhinia purpurea DC. Ex Walp. Fabaceae

Purple Bauhinia, Camel hoof tree (Eng.), Gulabi Kachnar (Hindi / Guj.), Dev Kanchan, Rakhta Kanchan (Beng.), Karath (Punj.), Boroda (Oriya), Atmatti (Mar.), Kanchanam (Tel.) Basavanapada (Kann.), Mandari (Tam.), Chuvanna Mandaram (Mal.).

A beautiful deciduous flowering tree with upright slightly crooked branches. The leaves are rounded, two lobed with an overall resemblance to camel hoofs. In northern India, leaf shedding occurs before March. In the high rainfall belt of southern India, plants are seldom leafless at flowering. Flowering starts in March and continues up to November. The fruits are flattened and carry five to six seeds. Trees are ideal for planting in residential avenues and small gardens. *Bauhinia variegata* with pale pink flowers is also grown. *Bauhinia racemosa* having white flowers occur in deciduous forests in India are referred to as Mountain Ebony.

Bignonia megapotamica Spreg. Bignoniaceae

Rio-Grande Trumpet flower

Plants are attractive evergreens with shiny leaves growing to a height of 12m. Native to Brazil, they were introduced into India as shade trees for gardens and avenues. The flowers are pale pink, typically trumpet shaped and are produced in small clusters during March-April.

Caesalpinia coriaria Willd. Fabaceae

Divi-divi

A small sized evergreen tree with an umbrella like canopy formed by low spreading branches. The leaves are dark green, pinnate and feathery in appearance. The flowers are greenish white, and the fruits are characteristically twisted. In its native habitat in central America, the fruits are used for tanning and dyeing. A popular shade tree planted around the borders of playgrounds and stadiums.

Callistemon lanceolatus (Sm.) Sweet Myrtaceae

The Bottle Brush tree

A slow growing medium sized tree attaining a height of 8m or more with characteristically drooping branches. The leaves are narrow and needle like and provide sparse shade beneath trees. The flowers with their showy red stamens are borne terminally on older branches giving them the appearance of bottlebrushes. Flowering extends from February to November. Ideal for home gardens as it can be trained to the stature of a small tree.

Cananga odorata Hook. Annonaceae

Ylang-ylang, Lang-Ilang

Trees are slender evergreens native to Burma and Malaya having gracefully arching branches. The scented greenish yellow flowers are produced in axillary clusters. The ovoid fruits are green and turn yellow on ripening. Flowers are used for extraction of perfumery oil.

Cassia fistula L. Fabaceae

Indian Laburnum (Eng.), Amaltas (Hindi), Bander lathi, Sonali, Sondal (Beng.), Sonaru (Assa.), Sunari (Oriya), Alash (Punj.), Garmalo (Guj.), Bhava (Mar.), Rela (Tel), Kakke (Kann.), Konnai (Tam.), Kani-konnai (Mal.).

Trees are the most spectacular among the yellow cassias grown in India. They are naturalized in deciduous forests up to an elevation of 1500m above

M.S.L. They attain heights of up to 10m and are deciduous to semi deciduous in different parts of the country. The slender branches have a drooping habit and carry long pinnate leaves with 4 to 8 pairs of leaflets. The bright yellow rounded blooms are produced in pendant panicles when trees are leafless in March – April or along with the new leaves from late April to May. The pods are long and cylindrical turning black on drying. The tree has the status of being the State Flower of Kerala.

Cassia javanica Linn. Fabaceae

Java Cassia

A medium sized tree attaining a height of 15m. The spreading branches and branchlets give trees an umbrella shaped canopy. The pinnate leaves have smooth and silky textured leaflets with rounded tips. Trees are completely deciduous in the dry tracts and semi-deciduous in humid regions of India. The pink flowers appear with the new leaves in May. The pods are cylindrical and contain flattened seeds.

A native of Java and Sumatra, trees were introduced into Indian gardens for their spectacular flowers.

Cassia nodosa Buch.- Ham. ex Roxb. Fabaceae

Flowering Cassia, Pink Cassia

Trees are relatively large deciduous evergreens and are similar in branching habit to *Cassia javanica.* The pink flower clusters arise from nodes when the trees are leafless from May to June. It is a particularly beautiful species suited for planting in large gardens and parks.

Cassia renigera Wall. Fabaceae

Burmese Pink Cassia

Trees are deciduous, attaining heights of up to 9m in Indian gardens. Upright branches bearing drooping branchlets give them an attractive canopy. The pinnate leaves bear feathery and downy leaflets. The clusters of rose-pink flowers arising from the nodes are lightly scented. Older flowers gradually fade to a white colour resulting in varying shades of pink and white flowers in the same cluster. Trees are ideal avenue trees for roadsides.

Casuarina equisetifolia L. Casuarinaceae

Beef wood tree, the Casuarina (Eng.), Jangli Saru (Hindi), Jhav (Beng.), Vilayati Saru (Guj.), Suru (Mar.), Sarugudu, Saruku, Chowka (Tel.), Chavukku (Kann.), Savukku (Tam.), Kattadi (Mal.).

A tall tree attaining heights of up to 30m with longitudinally exfoliating bark. The leaves are 6 to 8 jointed needles. Plants produce cones profusely from February to October. Trees are ideal for training as barrier hedges along garden boundaries. They have proven to be suitable for afforestation of sandy tracts and as salt resistant windbreaks along the seacoasts of South India. In *Casuarina* groves, the passage of wind through the branches and leaves produce a characteristic audible murmur.

Citharexylum quadrangulare Jacq. Verbenaceae

Fiddle-wood tree (Eng.), Paarijatham (Mal.).

Trees are upright, producing slender branches all around and attaining heights of up to 10m. Drooping racemes of small fragrant white flowers are produced in leaf axils towards the tips of branches. Plants remain evergreen in the humid tracts of peninsular India.

Couropita guianensis Hook. Lecythidaceae

Cannon-ball tree (Eng.), Topi gola (Hindi), Naga linga (Beng.), Kailaspathi (Mar.), Nagaalingam (Tam. / Mal.)

A tall erect tree attaining heights of up to 20m.The branches are pendant and leaves are crowded at the tips. The leaves are alternate, simple, obovate, glabrous and are shed once or twice a year. The flushes are light green in colour. The scented flowers are borne on long woody cauliflorous racemes and have a fascinating structure. The flower buds are rounded in shape. The sepals and petals are cupped and coloured in tints of pink and yellow. They surround the essential organs consisting of the gynoecium which resembles a "Shiva Lingam" over hung on one side by a snake's hood like structure formed by fused stamens. The divine significance attributed to these flowers makes this tree a specimen tree of gardens. The fruits are brown and cannon ball like and hang pendant close to the trunk. A native of tropical America, the tree was introduced for providing shade in public parks and gardens.

Crataeva nurvala Buch.-Ham. Capparidaceae

Caper tree, the Sacred Barna (Eng.), Barna (Hind / Punj.), Barun (Ass.), Barum (Beng.), Baruna (Oriya), Varno (Guj.), Haravarna, Vayavarna (Mar.), Mavilinga (Kann.) Maralingam (Tam.), Neeer mathalam (Mal.).

Native to deciduous forests, trees are distributed in Northern Kerala and attain heights up to 10m. They are planted near sacred places. The trifoliate leaves with ovate-lanceolate leaflets are shed in December- January. Small white flowers in dense clusters produced when leafless, make trees singularly beautiful. The flowers turn yellow with age and are shed before refoliation. Fruits are globular, woody and contain angular seeds with hard seed coats.

Cupressus spp. Cupressaceae

Cypress trees

Several species of evergreens with architectural canopy outline are grown as avenue trees in gardens located in the temperate belt of India. They include the mourning or weeping cypress (*Cupressus pendula*), the Kashmiri cypress (*C. cashmeriana*), the Monterey cypress (*C. macrocarpa*), the Mediterranean cypress (*C. sempervirens*), the Italian cypress (*C. fastigata*) and the Bhutan cypress (*C. nepalensis*).

Delonix regia Rafin. Caesalpinaceae

The Flamboyant tree (Eng.), Gul Mohr (Hindi / Mar. / Beng. / Guj.), Krishnachura, Radhachura, Rakhthachura (Ass. / Oriya), Bhima Sankesula (Tel.), Mayooram (Tam.), Alasipoomaram (Mal.).

Trees are large with stout trunks and a spreading canopy and attain heights of up to 15m.The bipinnate leaves give the canopy a fine feathery texture. The flowers are orange red to scarlet coloured. The trees are truly flamboyant when in flower giving the impression of the crown being ablaze. Leaf shedding occurs just before flowering. Occasionally flowers emerge with the new leaves. The fruits are long, flat and sword like. Trees are native to Madagascar.

Enterolobium saman Prain. Mimosaceae
Syn. *Samanea saman* Merrill., *Pithecolobium saman* Benth.

The Rain Tree (Eng.), Vilayati siris (Hindi, Guj., Mar.), Bilati siris, Koroi (Beng.), Nidraganneru (Tel.) Thoonga moonji (Tam.), Urakkam thoongi (Mal.).

Trees are large with thick and heavy spreading branches. The extensive canopy provides dense shade except during the short deciduous spell of December-January. Refoliation follows immediately in the humid tracts of India. Leaves are bipinnate and with a relatively large terminal pinna. Flowers are produced in small clusters and are pink or rarely white. Trees are native to Brazil and are popular along roadsides and playgrounds.

Erythrina variegata Fabaceae

Variegated Coral tree

Leaves are attractively variegated with patches of creamy white, pink and light green.

Ficus benghalensis Linn. Moraceae

The Banyan (Eng.), Bar (Hindi), Bat, Bar (Beng.), Borh (Punj.), Wad (Mar.), Aal (Tam.), Peyraal (Mal.).

Trees are large spreading and attain a grove like structure by virtue of their thick prop roots growing like supporting columns around the main stem. Birds relish fruits and the groves resound with their noisy chirping when fruits ripen. Suited for planting in large parks and gardens and for making Bonsai. However, the trees are revered as they are considered to be one of the Bodhi trees of other Bodhisattwas (previous incarnations of Buddha) and are cared for wherever they are grown.

Ficus comosa Roxb. Moraceae
Syn. *Ficus benjamina* L. var. *comosa*

Java Fig, Golden fig, Weeping fig (Eng.), Marri (Tel.), Aalmaram (Tam.), Ithi (Mal.)

A very ornamental evergreen tree that is native to and distributed in tropical Asia. The drooping branches carry ovate leaves that are glossy and leathery. The aerial roots hang down the branches and are thin and wiry. After flowering, small globose fruits shower the branches and ripen to a bright red colour. Birds relish them. A small to medium sized tree that can be pruned to suit small gardens.

Ficus elastica Roxb. Moraceae

India rubber plant (Eng.), Atha Bor (Beng.)

Trees are tall and attain heights of over 15m. The leaves are leathery and emerge covered in showy bracts which are coloured light pink to dark maroon. The mid rib may also be coloured similarly. Important cultivars include "Decora" and "Rubra". Leaves having variegations of pink, yellow and white are grouped under *Ficus elastica* var. *vareigata*. "Decora Schijvareana" is a popular cultivar. This species is more popular as ornamental foliage plants.

Ficus glomerata Roxb. Moraceae
Syn. *Ficus racemosa* Linn.

Gular (Hindi), Damur (Beng.), Umbar, (Mar.), Atti (Tel.), Athi (Kann. / Mal.)

Trees are tall, attaining heights of over about 30m.and have a smooth trunk. The leaves are ovate, oblong to lanceolate and characteristically get infested with galls. Cauliflorous, the fruits are produced in profusion. They ripen to a dark pinkish red colour and are shed like flowers, littering the canopy-basin. The young leaves and bark are used in indigenous medicine.

Ficus hispida Linn. Moraceae

Rough leaved fig, Devils fig (Eng.), Daduri, Katgulkar (Hindi), Dhed umbar (Guj.), Karoti (Mar.), Peyathi (Tam.), Kattathi, Peyathi (Mal.)

Trees are small sized and attain heights of about 10m. The trunk and branches are covered with small stiff hairs. The leaves are opposite, oblong, with an entire or toothed margin and a rough lower surface. Cauliflorous like most *Ficus* spp., the fruits become yellow in colour on ripening.

Ficus infectoria Roxb. Moraceae

Paku, Pilkhan (Hindi)

Trees are small in size. Leaves resemble those of *Ficus comosa* in shape but are bigger with well-marked venation. Tropical America is the center of origin of this tree.

Ficus religiosa Linn. Moraceae

The Peepal tree (Eng.), Peepal (Hindi), Asvatha (Beng.), Ravi, Ragi, Rai, Asvathamu (Tel), Drasi, Arasi (Tam.), Arasu, Arayaal (Mal.).

This is the Bodhi tree under which Gautama Buddha attained enlightenment. It is perhaps the most revered among the indigenous *Ficus* trees of India. Trees are large, attaining more than 30m height with tall straight trunks that branch higher up, creating a compact broadly rounded crown. The glossy heart shaped leaves have elongated tapering tips and they flutter in the slightest breeze. The entire crown resounds musically when windy. Trees are planted in temple compounds and large gardens for shade and resting places are constructed around them. *Ficus religiosa* var. *krishnae* called Makhan katori or Krishna Bar is a relatively small sized tree having cupped leaves that are cone-like at the basal end of the lamina. They are believed to have been used by Lord Krishna for storing stolen butter.

Ficus triangularis Moraceae

Angular-leaved ficus

Plants are relatively short in stature and are ideal for home gardens. The leaves are attractively angular. Shoots can be pruned and maintained like shrubs. They produce aerial roots sparsely.

Filicium decipiens Thw. Sapindaceae

The fern-leaved tree

Trees are medium sized, evergreen and attain heights of about 12m. The leaves are finely cut with winged petioles. This attractive ornamental foliage tree is native to South India and Sri Lanka.

Grevillea robusta A. Cunn. Proteaceae

Silver oak, Silk oak, Australian silk.

Trees are evergreen and grow up to 30m in height. The leaves are fern like and their under surfaces have a silvery sheen. The flowers are golden yellow and appear in April-May. The seedlings are grown as indoor plants.

Jaccaranda ovalifolia R. Ber. Bignoniaceae
Syn. *Jacaranda mimosaefolia* D. Don.

Jaccaranda (Eng.), Neeli Gul mohr (Hindi)

Trees are deciduous and attain a height of 8 to 9m. The open crown carries dark green leaves that are bipinnate. The flowers are blue, bell-shaped and are produced in clusters from February onwards in different parts of India. Trees are popular in gardens and are planted along avenues.

Lagestroemia flos-reginae Retz. Lythraceae

Queen's flower, Pride of India (Eng.), Jarul (Hindi), Jarool (Beng.), Ajhar (Assa.), Jarulo (Oriya), Taman (Mar.), Varagogu (Tel.), Maruvachall (Kann.), Kadali (Tam.), Manimaruthu (Mal.)

Trees are evergreen and grow up to a height of 20m. They are one of the most attractive trees of Indian gardens. The branches are short and sturdy and the relatively large leaves provide a dense canopy cover. The foliage is bright green and contrast beautifully with the pink to mauve coloured flowers. In peninsular India, the trees flower in April- May and August- September.

Melia azaderach Linn. Meliaceae

Bead tree, China tree, China berry, The Persian Lilac (Eng.), Dake, Drek, Bakain (Hindi), Ghora neem, Maha neem (Beng.), Pejri (Mar.), Mahanimbo (Oriya),Taraka vepa (Tel.), Malai vembu, Poo vembu (Tam.), Malaveppu (Mal.)

Trees are medium sized and attain heights of up to 8m. Seedlings are hardier than those of Neem but are damaged by frost. The leaves are dark green with finely cut margins. The canopy provides light shade and is ideal for residential avenues. The flowers are scented and lilac coloured. The fruits ripen to a yellow colour.

Michelia champaka Linn. Magnoliaceae

Golden Champa (Eng.), Champa (Hindi, Punj.), Phul chompa (Assa.), Sonar champa (Beng.), Pito champo (Guj.), Son champa (Mar.), Champakamu (Tel.), Sampige (Kann.), Shambaga (Tam.), Chempakam (Mal.)

These elegant evergreens are indigenous to India and attain heights of about 10 to 12m under cultivation. In the Himalayan foothills, trees attain heights of up to 30m. The trunk grows straight and slender branches are spread all around giving the canopy a broadly pyramidal shape. The leaves are smooth surfaced and ovate-lanceolate in shape. The young shoot buds are protected by scaly stipules. The flowers are light orange-yellow to pale yellow or creamy white in colour and have an intoxicating fragrance. They are produced singly in leaf axils. For earliness in flowering, trees are propagated by layering. The fruits are woody, ovoid in shape and have angular seeds.

Millingtonia hortensis Linn. Bignoniaceae

Indian cork tree, the Jasmine tree (Eng.), Akash Neem, Neem chameli (Hindi) Akas Neem Phulari (Beng.), Bakeni, Reali, Mach- mach, Sita-hara, (Oriya), Akas Nimb, Kavla Nimb (Mar.), Kavuki (Tel.), Beratu (Kann.), Mara Malli, Kattu Malli (Tam.), Mara Mulla (Mal.)

Trees are tall stately and evergreen, attaining heights of up to 20m. The gracefully drooping branches carry small and multifid leaves. The pure white trumpet shaped flowers have long corolla tubes and are delicately scented. They are produced in profusion during monsoons. Flowers open at night and are shed by morning, carpeting the ground below the trees. These were favored flowering trees of Indian gardens of the previous century.

Mimusops elengi Linn. Sapotaceae

Mulsari (Hindi), Bakul (Beng.), Bokolo (Oriya), Ovalli (Mar.), Pagada (Tel./ Kann.), Vagulam (Tam.), Ilenji (Mal.)

Trees are tall growing evergreens indigenous to India and attain heights of more than 15m. They have attractive glabrous leaves. The creamy white flowers are small and fragrant. They fade to a light brown colour and are shed quickly. Seedlings are slow growing. Layered plants have a bushy stature and are used as foliage plants.

Muntingia calabura Linn. Verbenaceae

Bird cherry

Trees are small sized evergreens having horizontally spreading branches that give the canopy an umbrella shape. The leaves are ovate oblique, with a rough surface texture. The flowers are white and the green berries ripen to a light yellow colour. Birds disseminate the seeds widely. They germinate readily like the seeds of the Peepal in crevices of cemented rooftops.

Nyctanthes arbor-tristis Linn. Oleaceae

Har singar (Hindi), Shephalika (Beng.), Parijat (Guj. / Mar.), Paarijaatamu (Tel.), Pavizha-Malli (Mal.)

Indigenous to India, these are small to medium sized trees that have a grayish green bark. The young branches droop downwards and are quadrangular in outline. The leaves are ovate and have widely serrate margins. Flowers are borne in terminal leaf axils from August to December. They are scented and have a bright orange coloured corolla tube terminating in free lobes that are white. Flowers open at night and are shed by morning. They were formerly used for dyeing silk thread and fabric. The fruit is a flat capsule with two seeds.

Peltophorum ferrugineum Benth. Fabaceae
Syn. *Peltophorum pterocarpum* Becker

Copper pod tree (Eng.), Konda chinta (Tel.), Ivalvagai (Tam.), Chara konna (Mal.)

Trees are large, with a spreading crown carrying dark green pinnate leaves. The flowers are golden yellow and borne terminally in pyramidal clusters during March to May. Mature fruits are coppery red in colour. One of the most popularly grown avenue trees of Indian highways, they are widely distributed in South East Asia.

Pisonia alba Nyctaginaceae

Cabbage tree, Lettuce tree, Pisonia

These are small evergreen trees native to the Andaman Islands. They are grown as foliage trees in tropical gardens for their attractive light green to yellowish green leaves. The leaves are broadly ovate and have serrate margins. They have medicinal uses. The young leaves are used as a vegetable in India.

Pongamia glabra Fabaceae

Pongam, Ungu (Mal.)

These are evergreens with a dense cover of drooping branches bearing glabrous ovate leaves. The flowers small, pinkish white and are produced in drooping inflorescences. The seeds are the source of Pongam oil are also being used for preparing bio-insecticides.

Plumeria acutifolia Poir. Apocynaceae
Syn. *Plumeria acuminata*

The Pagoda tree, The Frangipani (Eng.), Chameli, Gul-i-chin (Hindi), Dalanaphula (Beng.), Golachi (Oriya), Khair champa (Mar.), Arhattaganneru (Tel.), Kadusampage (Kann.), Ilattarali (Tam.), Arali, Ezha chembakam (Mal.)

Trees are small to medium sized with an evenly spreading crown. The bark is silvery gray in colour. Leaves are broadly lanceolate, tapering at both ends. The flowers are fragrant and coloured in varying shades of pink and yellow. They are produced from March- April to October. Trees are native to Mexico and Guatemala. They are conventionally planted near pagodas and mandapams.

Plumeria alba Linn. Apocynaceae

The Pagoda tree, White Frangipani (Eng.), Veyyivarahalu (Tel.), Peru (Tam.)

Trees are short in stature with dark green leaves. They are native to India. The scented white flowers are larger than the flowers of other species and have reflexed petals.

Plumeria rubra Linn. Apocynaceae

Lal Champa

Trees resemble *Plumeria alba* in stature. The flowers are attractive and pinkish-red in colour. They are produced in loosely hanging clusters during March April and July- October.

Polyalthia longifolia Benth. & Hook. Annonaceae

The mast tree (Eng.), Asokam, Deodar (Hindi), Debdaru (Beng.), Aranamaram (Mal.)

Trees are tall and evergreen. They have evenly spreading branches giving the crown a pyramidal outline. The leaves are glossy and have wavy margins. Greenish flowers are produced in March- April. *Polyalthia longifilia* var. *pendula* has pendant branches that hang close to the trunk, covering it and is more popular for planting in gardens and along central avenues of double-lane roads and highways.

Ravenala madagascariensis J. F. Gmel. Musaceae

Travellers Palm, Travellers tree (Eng.), Khajur Pankhi (Hindi), Panthapadup (Beng.)

Trees have fan like leaves held on either sides of the main shoot tip in a single plane. The leaves resemble the leaves of bananas. However the trees have a woody trunk resembling that of palms. They are planted as specimen plants in gardens for these unique features.

Saraca asoka (Roxb.) De Wilde Fabaceae
Syn. *Saraca indica* Linn.

The Asoka tree (Eng.), Ashok, Sita Asoka (Hindi / Beng.), Ashopalava (Guj.), Osoko (Oriya), Asoka, Jasundi (Mar.), Asokamu (Tel.), Asoka (Kann.), Asogam (Tam.) Asokam, Hemapushpam (Mal.)

Trees are evergreen, short in stature and the trunk has a warty surface. Leaves are pari-pinnate, with oblong-lanceolate leaflets. Flowers are orange red in colour and are produced in dense clusters. The fruits are compressed and contain 2 to 3 seeds. Trees are adapted to the shady conditions of evergreen forests and occur in groups near streams.

Spathodea campanulata Beauv. Bignoniaceae

The African tulip tree, Fountain tree, Scarlet bell tree (Eng.), Patadi (Tel.), Nirukavi (Kann.)

Trees are large, evergreen and attain heights of 25-30m. The branches are short, giving trees a compact crown. Leaves are pinnate, ovate-lanceolate and consist of 9-19 pairs of leaflets in addition to the odd terminal one. Flowers are orange-red coloured and are produced in terminal erect panicles from October to December. They carry nectar and birds largely effect pollination. Fruits are erect pointed pods. Trees are native to central Africa and were introduced into India for planting in gardens and along roadsides. At present it is widely distributed in the lower slopes of the Western Ghats.

Swietenia mahagonii Jacq. Meliaceae

Mahogany (Eng.), Abnus (Hindi), Mehagoni (Beng.) Mahagony (Mal.)

Trees are large, deciduous, native to central and South America. They were introduced into India for their silvicultural importance. The glossy leaves and extensive canopy provide shade almost throughout the year except during the brief period of leaf fall in December-January. The fruits are woody, ovate to rounded in shape and more conspicuous than the small white flowers. They dehisce on drying releasing winged seeds in large numbers that are dispersed by wind.

Tabebuia spectabilis (Planch. & Linden) G. Nicholson Bignoniaceae
Syn *Tabebuia chrysantha* (Jacq.) Nichols.

Golden Trumpet tree (Eng.), Basant Rani (Hindi)

Trees are native to South America. They are small in size and have a rough textured and fissured bark. The older branches have a crooked appearance, which enhance the aged and attractive look of the trees. The mature leaves have a grayish green colour. Golden yellow flowers are produced in profusion during March-April. *Tabebuia rosea* D.C., native to Mexico is the Pink Trumpet tree and is grown in the less humid tracts of India. *Tabebuia aurea* Benth. & Hook. f. ex S. Moore is the Silver Tabebuia or Carribean Trumpet tree.

Tamarindus indica Linn. Caesalpinaceae

Amli, Ambli (Hindi / Guj. / Beng.), Tentul (Beng.), Chincha (Mar.), Chintachettu (Tel.), Puli (Tam.), Vaalan puli (Mal.)

A large and long-lived tree native to India and Africa, it is one of the more popular shade trees of roadsides in South India. Leaves are compound with a chanelled axis and carry 10 –20 pairs of leaflets. The mature leaves are jade green in colour. The flowers are small and yellowish in colour with pink stripes on petals. They are produced in May-June. The fruits are slightly curved and greenish brown in colour. The seeds are shiny coffee brown in colour, square in outline and slightly flattened. The fruits are harvested when tree-ripe and are the source of an acidic pulp having culinary and medicinal uses.

Terminalia arjuna Wight & Arn. Combretaceae

Arjun (Hindi / Beng.), Yermaddi (Tel.), Kulamaruthu, Vellamaruthu (Tam.), Neermaruthu (Mal.)

Plants are semi-deciduous foliage trees grown for shade in gardens and along road sides. the leaves hang gracefully on drooping branches.The flowers are creamy white and inconspicuous.

Terminalia catappa Linn. Combretaceae
Syn. *Terminalia badamia*

Tropical almond, Indian almond, Bengal almond, Country badam (Eng.), Badam, Deshi badam, (Hindi), Bangla badam, Baksa badam (Beng.), Badamo (Oriya), Jungli badam (Mar.), Ingudi, Nadu badamu (Tel.), Badamalili, Nadu badami (Kann.), Amandi, Nattuvadum (Tam.), Thalli thenga (Mal.)

Trees are tall, deciduous and have a straight trunk and whorled branches that spread horizontally. Leaves are large and found tufted at the end of branches. They acquire a red colour before shedding in December- January. Flowers are small, white and are produced in slender spikes. The fruits are egg-shaped and flattened along the edges like a ridge. They turn yellowish red on ripening. Bats relish the outer pulp of the fibrous fruits. Trees are grown for shade and for their canopy architecture in gardens and roadsides.

Thespesia populnea Soland. Malvaceae

Portia tree (Eng.), Paras pipal (Hindi), Bendi, Parasapiplo (Guj.), Parash, Porassh (Beng.), Bhendi (Mar.), Gangaravi (Tel.), Kallal, Poovarasu (Tam.), Cheelanthi, Poovaras (Mal.)

Trees are small to medium sized, evergreen and have closely spaced branches and a rounded canopy. The leaves are heart shaped with acuminate tips. The flowers are large, yellow, cupped and have a maroon-red colour on the inner surface of the petals at the basal end. They are salt tolerant and also withstand waterlogged situations are hence planted in coastal regions and in frequently inundated areas. Trees were popular shade tree of roadsides during the previous millennium in South India.

Platycladus orientalis (L.) Franco Cupressaceae

Chinese Arbor vitae / Biota

Plants are similar to *Thuja occidentalis* (Arbor Vitae) and were previously named *Thuja orientalis* L. The finding of a distant relationship only with the former has resulted in revision of its systematic position and renaming. Small sized trees of this conifer are grown in Indian gardens. They are also planted along avenues and in pots for their architectural shape and curiously flattened and angular branches carrying scale like leaves that are not aromatic unlike the 'Arbor vitae'. With ageing, the trunk becomes twisted and the bark sheds in longitudinal strips.

PALMS AND CYCADS

Palms belonging to Arecaceae and Cycads belonging to Cycadaceae having ornamental value are grown in gardens against the backdrop of an expansive lawn or along avenues and as indoor plants. They attain the stature of trees and have a compact crown of evergreen leaves. Adapted to warm humid tropical situations, they are hardy and are propagated by seeds. The suckering types can be propagated by separation of suckers from clumps.

PALMS

Ornamental palms are grouped into two, based on the extent to which their leaflets are united. Those in which the individual pinnae are free are called the **Feather leaved Palms** and those in which the pinnae are partially to completely united are called the **Fan leaved Palms**. The popularly grown ornamental palms of our gardens belonging to both these groups are listed below: -

FEATHER LEAVED PALMS

Caryota mitis Lour.

Clustering Fish-tail palm

These are quick growing and attain heights of up to 7m. The stems are stout and ringed. The leaves are 1-3m long and have wedge shaped leaflets that resemble tail fins of fish. They produce suckers and are indigenous to Burma and Indonesia.

Chrysalidocarpus lutescens H. Wendl.

Golden Cane palm, Parlour palm

These are native to the Bourbon Islands. The stems are slender and golden yellow in colour. These palms are grown in large clumps. The leaves become beautifully arched as palms mature. They are excellent for planting in lawns and indoors.

Cyrtostachys renda
Syn. *Cyrtostachys lakka*

Ornamental Sealing-wax palm

These are tall graceful palms, native to Borneo. The sheathing petiole and midrib of the leaves are bright red in colour. They are very ornamental when young. Tall stems are cut back to maintain clumps attractive.

Phoenix canariensis

Canary Date palm

These are tall stately palms having spiny leaves that are held straight without an arch.

Phoenix roebelenii

Pygmy Date palm

Plants have finely divided arching pinnae. They are short, slow growing and are suited for sunny indoor areas.

Phoenix sylvestris Roxb.

Wild Date palm

This is a slow growing palm with feathery dark green leaflets. It is suited for growing indoors in the juvenile stage.

Roystonia regia C.F.Cook

The Royal palm, the Bottle palm

Palms are majestic and stately attaining heights of up to 20m. The trunk swells gracefully at the middle as plants age. Slow growing and suited for avenue planting. Leaves are feathery and 2-4m long. The sheathing petiole fully encloses the trunk. They are native to the West Indies

FAN LEAVED PALMS

Licuala grandis H. Wendl.

The Table palm

These are short in stature and are very popular in gardens. The leaves have ruffled or pleated lobes and the margins of leaflets are united and have toothed tips. The petiole is slender and spiny. Grown as an indoor plant too.

Livistona chinensis R. Br.

Chinese fan palm

These are hardy palms that attain heights of up to 9m. They are extensively grown in gardens. The leaves are broad and deeply forked and have leaflets that are partially free. The crown is compact and globular.

Rhapis excelsa
Syn. *Rhapis flabelliformis*

Ground Rattan palm, Rathan cane palm Large Lady palm (Eng.), Kannonochiku (Jap.)

These have thin bamboo like stems that attain heights of up to 3.5m. They grow in clusters. Persistent fibrous leaf sheaths cover the stem. The leaves are glossy, dark green and cut into several segments. Petioles are toothed. They are native to China.

CYCADS

Cycas circinalis Linn.

Fern Palm, Palm Cycad

These have cylindrical stems that attain heights of up to 4.5m. The stems are initially erect and tend to recline with age and are covered by leaf scars. Branching is frequently observed in old plants.The crown carries a rosette of glossy pinnate leaves. Plants may be male or female. They are distributed in India, Madagascar and New Guinea.

Cycas revoluta Thunb.

Ornamental Sago palms, Dwarf Cycas

These cycads are dwarf in stature, attaining heights of up to 3m. The main stem branches occasionally above the ground level. The leaves are dark green and have short and stiff leaflets. Plants are native to Java and are found naturalized up to Southern Japan.

Zamia furfuracea

Cone-leaved Cycas

Plants are similar to *Cycas circinalis* in habit. They are slow growing and thick-stemmed and the main stem remains subterranean for a long time. The leaves have stout mid ribs and the leaflets are thick, leathery, dull green, angular and broadly pointed towards the tip. Suckers are produced from the base in its natural habitat. In gardens, plants thrive near pool sides.

CONIFEROUS TREES AND DECIDUOUS TREES FOR TEMPERATE CLIMATIC SITUATIONS

In the hill zones of the tropics and in regions characterized by warm to cool temperate climatic conditions, evergreen and deciduous trees are grown in gardens and landscapes. Conifers are mainly valued for their architectural features and evergreen nature while deciduous trees that thrive in these situations transform the landscape from late summer through autumn with the vivid colours attained by their foliage.

CONIFEROUS TREES

These are mostly large evergreens having columnar to pyramidal structure and needle like or scale like leaves and usually produce seed bearing cones. They are mostly confined to the northern hemisphere and are among the more primitive of seed producing plants.

Coniferous trees grown in gardens include the following: -

Abies sp. Mill. Pinaceae

Fir trees

Thisis a genus of evergreen trees having large to small conical crowns and are distributed widely in the temperate regions. Firs are very popular as Christmas trees. *Abies grandis* the Giant Fir, *Abies cephalonica* the Grecian Fir and *Abies pinsapo* the Spanish Fir are large in size and *Abies koreana* the Korean Fir is a small tree and *Abies balsamea* the Balsam Fir and *Abies concolor* the white Fir are compact to spreading shrubs.

Agathis australis Araucariaceae

Dammar Pine, Kauri Pine, New Zealand Kauri

These are native to the southern hemisphere and have conical crowns when young and develop large trunks and rounded canopies when mature.

Araucaria araucana Araucariaceae

Monkey's Puzzle tree

Araucaria Juss. is an ancient genus of evergreens native to the southern hemisphere. They are large trees with erect stems and horizontal branches carrying condensed and curved needle like leaves. The elongated conical to pyramidal shape of the trees and dark green foliage make them popular in formal plantings. The Monkey's Puzzle is the national tree of Chile. *Araucaria angustifolia* the Parana Pine and *Araucaria cunninghami* the Hoop Pine are also popular in South American gardens. A few species are also grown in Indian gardens - for details see under '**Flowering and Foliage Trees**'.

Casuarina equisetifolia (See under '**Flowering and Foliage Trees**')

Cedrus deodara (Roxb.) G. Don Pinaceae

Deodar, Himalayan Cedar (Eng.) Devadaru (Mal.), Deva Tharu (San.)

Native to the Himalayas, stands of these tall growing large trees with broadly pyramidal crowns and slightly pendulous branches are a majestic sight of the slopes and valleys and are called the 'Trees of Gods'. The young leaves are blue-green and have a fine texture.

Cryptomeria japonica (L.f.) D. Don Cupressaceae

Japanese Cedar, Sugi (Jap.)

This is the national tree of Japan. Plants have large trunks and spirally arranged needle like leaves. The wood is scented and extensively used for buildings. A traditional tree of Japanese gardens, plants with shrub like habit and having juvenile leaves alone are also grown.

Cupressus L. Cupressaceae

Cypress trees

This genus has evergreen trees or arboreal shrubs native to the warm temperate regions of the northern hemisphere. They are decorative plants of traditional English and Mughal gardens. Several species having an architectural canopy outline are grown as avenue trees in gardens located in the temperate belt of India. They include the mourning or weeping cypress (*Cupressus pendula*) and the Kashmiri cypress (*Cupressus cashmeriana*). Others having ornamental value are the Monterey cypress (*Cupressus macrocarpa*), the Mediterranean cypress (*Cupressus sempervirens*), the Italian cypress (*Cupressus fastigata*) and the Bhutan cypress (*Cupressus nepalensis*). The leaves are scale like and arranged in pairs.

Ginkgo biloba L. Ginkgoaceae

Maidenhair tree

Plants are long-lived deciduous tall growing trees distributed in Eastern China. Considered to be living fossils related to gymnosperms among present day plants, they have fan shaped leaves with notched edges. The veins are radiating and branch dichotomously without forming a network.

Juniperus L. Cupressaceae

Junipers

This is a genus of coniferous trees and arboreal shrubs distributed in the northern hemisphere. *Juniperus chinensis* L. the Chinese Juniper and *Juniperus communis* the Common Juniper are popular garden plants of Indian gardens too. The berries are traditionally used for flavouring food in their centers of origin.

Larix decidua Mill. Pinaceae

European Larch

These are deciduous trees with open crowns and a silvery-gray to grayish-brown bark. The leaves are flattened and sessile.

Metasequoia glyptostroboides Hu & Cheng Cupressaceae

Dawn Redwood

Plants are large, fast growing deciduous conifers that thrive in waterlogged soils. They are endangered in the wild and are protected in national parks in N. America.

Picea abies Pinaceae

European Spruce

These belong to a genus of coniferous evergreens distributed in the northern temperate and artic regions. They are long-lived and have a conical structure with whorled branches and are extensively used as Christmas trees.

Pinus L. Pinaceae

Pine

This is a large genus of coniferous trees native to the northern hemisphere. They are the chief timber trees of the temperate belt. The mature leaves are found in whorl like spirals and are called needles. The seed bearing cones are used ornamentally. Pines are considered to be invasive and large pine forests tend to reduce the natural plant diversity of regions. Species diversity of pines is high in Central America.

Podocarpus Labill. Podocarpaceae

Fern Pine, Pine

This is a genus of monoecious or dioecious evergreen trees or arboreal shrubs of ancient origin. They are found in almost all the continents including Antarctica. The leaves have a distinct midrib and are spirally arranged on the shoots. Foliage is used in fresh floral arrangements. The female cones at maturity become fleshy and reddish purple in colour and are edible. *Podocarpus macrophyllus* the Buddhist Pine is an arboreal shrub. *Podocarpus nerifolia* is found to thrive in the humid belt of South India. The leaves are used decoratively.

Sequoia sempervirens Cupressaceae

Coast Redwood, California Redwood

Plants are evergreens and are the tallest among the trees of the world. Together with two other species they belong to the only living genus of redwoods. The freshly cut wood of these plants is reddish brown in colour. Plant specimens are maintained in bio reserves and national parks in N. America.

Sequoiadendron giganteum (Lindl.) J. Buchh. Cupressaceae

Giant Sequoia, Sierra Redwood, Wellingtonia

These are the world's largest trees and occur naturally in California in scattered groves that are protected. They are evergreen with awl shaped leaves that are arranged spirally.

Metasequoia glyptostroboides Hu & Cheng Cupressaceae

Dawn Redwood

Plants are fast growing deciduous conifers that can thrive in waterlogged soils. Trees attain a height of 30m and a diameter of 1.3m. They are critically endangered in the wild.

Taxodium Rich. Cupressaceae

Cypress trees

This is a genus of deciduous to evergreen trees endemic to the eastern coast of North America. A number of species are distributed in swampy areas and highlands. They include the Pond Cypress *Taxodium ascendens* and the Montezuma Cypress *Taxodium mucronatum.*

Taxus baccata Taxaceae

European Yew

These are slow growing long-lived coniferous trees that have spirally arranged lanceolate leaves held distichously. They are the conventional hedge plants of European gardens. Seed cones have a modified scale that develops into an aril enclosing the seeds. These are bright red and are called yew berries and are eaten by birds. All plant parts except the aril are poisonous. The Himalayan Yew is *Taxus wallichiana.* The only tropical species is the Sumatran Yew *Taxus sumatrana.*

Thuja occidentalis Cupressaceae

Arbor vitae (Tree of Life), Eastern *Arbor vitae,* Northern White Cedar, Thuja, Thuya

These are small evergreen trees having fan like flattened branches held on a single plane.

The foliage consists of scale like leaves and is characteristically scented. They are garden favourites in Europe and North America.

Tsuga Carr. Pinaceae

Hemlock

This is a genus of evergreens adapted to cool and moist areas of temperate regions. Many species are very cold hardy. The plants have irregular to conical crowns carrying flattened branched with drooping ends. The leaves are aromatic and their aroma resembles that of a herb poison called hemlock obtained from an unrelated plant. *Tsuga dumosa* the Himalayan Hemlock has pendulous pubescent branches.

DECIDUOUS TREES

These are the trees most commonly found in large gardens of the temperate regions. They are grown for shade, for the beauty of their foliage and flowers and for the bright colours the leaves attain in the autumn season or fall. A large number of these are distributed in the northern hemisphere. Most of them are cold hardy but those that cannot tolerate spring frost that harms new buds and leaves are usually seen confined to warmer areas.

Popularly grown ornamental **deciduous trees** include the following: -

Acer L. Aceraceae

Maple

This is a genus of trees and shrubs native to the northern hemisphere. The species distributed in Asia and the Mediterranean regions are mostly evergreen and those distributed further north are deciduous. The fruit is a samara and is referred to as 'maple key'. Around 15 species including *Acer caesium* are found in the Himalayas. The Sugar Maple is *Acer saccharum*, the source of maple syrup, which is the sap tapped from the bark of mature trees. The foliage of Maple trees attains brilliant yellow and red colours before leaf fall.

Betula L. Betulaceae

Birch (Eng.), Bhoorj (Hindi)

This is a genus of trees having bark coloured red, black, white, yellow or silver and marked with horizontal lenticels that separate in thin papery plates. These, referred to as 'Bhoorj Patra' were used in ancient times in India for writing. Birch is the national tree of Russia. The Silver Birch is the national tree of Finland.

Carya glabra Juglandaceae

Pecan, Pignut Hickory, Sweet Pignut

These have ornamental foliage. The leaves are yellowish green and turn reddish in Autumn. The flowers are drooping catkins. The seeds are edible.

Cercis canadensis Fabaceae

Eastern Redbud / Judas Tree

These are flowering trees native to N. America and are popular in gardens. A number of species are grown in C. America, Europe and the Mediterranean region.

Cornus florida Cornaceae

Flowering Dogwood

These are popular shade adapted garden trees of N. America and are grown for their brilliant red autumn foliage and clusters of white to pink floriferous bracts that cover plants during the flowering season.

Cotoneaster Medik. Rosaceae

Cotoneaster

The genus is native to temperate regions and is grown in gardens for their white to rosy pink flowers and showy bright red berries. The Himalayan Tree Cotoneaster is grown in parks and large gardens of N. India. Other species grown are *Cotoneaster khasiensis* having rose-pink flowers and *Cotoneaster acuminata,* which is used for making walking sticks.

Fagus sylvatica Fagaceae

European Beech

These are native to the mountainous regions of Europe. They occur in dense stands crowding out other vegetation and allowing very little light into their interiors. The leaves are simple with crenulate margins. The flowers are produced in small catkins and the fruits called beechnuts are edible.

Fraxinus excelsior Oleaceae

Common Ash, European Ash

Plants are ornamental foliage trees with drooping branches that are popularly grown in gardens. The flowering Ash is *Fraxinus ornus.*

Laburnum anagyroides Fabaceae

Laburnum, Golden Chain tree

Plants are small trees native to Europe. The flowers are golden yellow and densely packed in pendulous racemes of up to 20cm length. The resemblance of these flowers to those of *Cassia fistula* of the same family has resulted in the common name 'Indian Laburnum' of the latter. All the plant parts are poisonous.

Liquidambar orientalis Altingiaceae

Oriental Sweet Gum (Eng.), Silaras (Hin.)

These were introduced into Indian gardens during the colonial period. The leaves are palmately lobed and yellow flowers are produced in globose heads. A resin named storax produced from the seeds has perfumery use.

Magnolia grandiflora Magnoliaceae

Southern Magnolia, Bull Bay

These are small ornamental trees native to N. America and are widely grown in temperate zone gardens. The tepals are showy, fragrant and white to pink in colour. *Magnolia campbelli* is native to E. Himalayas and Manipur. *Magnolia obovata* is the Japanese Big leaf Magnolia and *Magnolia liliflora* is the Lily Magnolia or the Tulip Magnolia native to China.

Platanus orientalis Linn. Platanaceae

Chinar, Oriental Plane tree, Sycamore, Buttonwood

These are trees that are often avenue planted along highways. The leaves are lobed and have serrate margins. The flowers are globose balls.

Populus Linn. Salicaceae

Aspen, Cottonwood, Poplar

This is a genus that has more than 25 species distributed in the northern hemisphere. They are differentiated into three groups and referred to by three common names- aspens, cottonwoods and poplars. Aspens are native to

Europe and N. America. They have leaves with flat petioles that tremble in the slightest breeze. The Common Aspen or Trembling Aspen of Europe is *Populus tremula* and the Quaking Aspen native to N. America is *Populus tremuloides*. Cottonwood trees are native to the wetlands of N. America. The Swamp Cottonwood is *Populus heterophylla*. Poplars are native to S. Europe and Asia. The White Poplar is *Populus alba* and the Black Poplar is *Populus nigra*.

Prunus Linn. Rosaceae

Flowering Cherry

These are cherry trees having multi-species parentage. They are popularly avenue planted as flowering trees.

Quercus L. Fagaceae

Oak

This is a large genus with more than 300 species of deciduous to evergreen trees attaining majestic proportions in size. They are native to the northern hemisphere. The leaves are characteristically lobed. The fruit is an edible nut called an acorn. Plants are components of large gardens and important sources of timber. Oak trees are the national trees of England, Wales, France, Germany, Estonia, Lithuania, Poland, Serbia and the United States of America. The English Oak is *Quercus robur*.

Salix babylonica Salicaceae

The Willow

Plants are important foliage trees of gardens of the temperate regions. The spreading branches have drooping tips. The Weeping Willow is *Salix chrysocoma*.

Ulmus hollandica Ulmaceae

Elm, European Elm

These are important foliage trees of European gardens. The branches have a weeping habit. *Ulmus wallichiana* is the Himalayan Elm

PROPAGATION AND PLANTING OF TREES

In the tropics, trees are to be selected based on their suitability to growing conditions. Seed propagation is resorted to for ensuring long life and the characteristic structure and branching habit of species. In dicotyledonous trees with dimorphic branching, vegetative propagation of orthotropic shoots will maintain plant architecture as in seed propagated plants while plageotropic shoots will not. Most of the leguminous tree seeds have dormancy and require scarification for quicker germination. Unscarified seeds take 2-3 months to germinate as their dormancy is overcome in nature over a period of one to three months.

Hardened one to three years old nursery plants are the most suitable for planting. Planting should preferably coincide with the monsoons for quicker acclimatization of plants. Pits of 1 cubic meter size are to be taken and if possible, left to weather for 2 to 3 weeks after partially filling them with 10-15 kilograms of organic manure. For row planting, small sized trees are spaced 5m apart while medium to large sized trees are spaced 9-10m apart. For grove planting, depending upon the species selected, a closer spacing is to be adopted.

AFTER CARE OF TREES

After planting, trees are to be staked and those planted along roadsides are to be protected with a tree guard. Except for avenue planted conifers and pines, tree trunks are generally left bare of branches up to a height of at least 2.5m. Lopping of shoot suckers (orthotropic shoots arising from near the base) and pruning of excessive lateral growth is also to be done during the initial years. Pruning of heavy branches is to be done with a suitable implement that can make a clean cut without unnecessary wounding of tree bark.

The life span of trees varies. Trees like the Redwoods, the immortal Banyans and several others have survived climatic and environmental variations over several millennia. However most of the ornamental trees of gardens show decline in vigour and re-growth of branches with ageing. Old trees are to be examined for symptoms of progressive decay. Those in which the trunks have become hollow and partially diseased are to be removed quickly. Under-planting can be done well in advance of felling old trees.

contd. colour plates

Couropita guianensis

Enterolobium saman

Nyctanthes arbor-tristis

Peltophorum ferrugineum

Saraca asoka

Tabebuia spectabilis

Bauhinia purpurea

Cananga odorata

Cassia nodosa

Citharexylum quadrangulare

Muntingia calabura

Plumeria alba

Amoora rohituka

Araucaria cookii

Bignonia megapotamica

Casuarina equisetifolia

Filicium decipiens

Podocarpus nerifolia

Ficus comosa

Ficus elastica

Ficus religiosa

Ficus repens

Ficus retusa

Ficus triangularis

Caryota mitis

Rhapis excelsa

Cyrtostachys renda

Licuala grandis

Livistona chinensis

Chrysalidocarpus lutescens

Chapter 2

SHRUBS : IMPORTANCE AND USES

Shrubs are the most diverse groups of ornamentals grown in tropical gardens. In habit they are hardy perennials having low spreading woody branches. They vary widely in stature and include dwarf bushes to those attaining the stature of small trees.

IMPORTANCE AND USES OF SHRUBS

Shrubs are quick growing and many of them flower continuously in the humid tropics. Their variability in flower colours and foliage characteristics and their easiness in propagation by vegetative means make them garden favourites. The various uses to which shrubs are put to in gardens are detailed below: -

In A Shrubbery : Grouping of shrubs in a shrubbery is done to display the attractiveness of their flowers and foliage. In temperate situations shrubs grown in shrubberies include Rhododendron, Peony, Fuchsia, Buddleia, and Camellia while in the tropics and subtropics, the range of shrubs available for planting are very wide.

As Focal Point Plants : Shrubs that are ever flowering like Hamelia and Hibiscus or those that have spectacular flowers such as Mussaenda, Neerium and Ixora are grown as focal point plants in small gardens. They are also grouped for mass effect in large gardens.

In Mixed Borders : Shrubs are grown to provide permanent colour with flowers and foliage and as the perennial component alongside annuals in mixed borders.

For Hedge Making : Shrubs that have small leaves and closely spaced branches, which can be frequently pruned to get a 'fine' surface texture, make excellent foliage hedges.

Examples include *Phyllanthus rosea, Randia maculata, Murraya exotica* and *Duranta plumieri*. Shrubs that combine the above foliage characteristics with attractive flowers are grown as flowering hedges. They include *Ixora chinensis, Barleria cristata* and *Crossandra*.

For Topiary : Shrubs that are grown, as foliage hedges are ideal for topiary. Depending upon the desired size of the topiary, bushy shrubs like *Phyllanthus rosea* to arboreal shrubs like *Murraya exotica* can be selected.

For Foundation Planting : Low-growing shrubs can be grown as Foundation-covers to mask the visible basement area of buildings. They also serve as transition plantings between the indoor and outdoor garden areas.

CLASSIFICATION OF SHRUBS

Shrubs are classified based on criteria such as differences in stature, light requirement, ornamental nature and method of use in gardens. These groupings are useful in selecting shrubs of different sizes for different situations and locations and for different kinds of uses in gardens.

CLASSIFICATION BASED ON STATURE

Based on the maximum height, spread and branching habit when grown in garden situations, shrubs are classified into **Arboreal shrubs**, **Bushy shrubs** and **Sub-shrubs**. Arboreal or tree-like shrubs are more than 2.5 to 3m tall and attain the woody nature and branching habit of small trees. Examples include *Murraya exotica, Ixora acuminata, Hamelia patens* and *Tabernaemontana indica*. Bushy shrubs attain a maximum height of 1 to 2m and an equally wide crown spread. *Hibiscus rosasinensis, Ixora singaporensis, Gardenia jasminoides* and *Tabernaemontana coronaria* belong to the larger sized category in this group. Bushy shrubs, which attain a maximum height of 1 to 1.5m, include *Barleria cristata, Duranta plumieri, Malaviscus arboreus* and *Cassia biflora*.

Sub-shrubs are low growing or dwarf shrubs attaining a maximum height below 1m. They include *Cuphea hyssopifolia, Galphimia gracilis* and *Jacobinea carnea.*

CLASSIFICATION BASED ON ORNAMENTAL NATURE

Shrubs can broadly be divided based on their ornamental nature into **Flowering shrubs** and **Foliage shrubs**. Shrubs, which have greater ornamental value due to their attractive flowers, are called **Flowering shrubs**. While those that have attractively coloured or variegated foliage are called **Foliage shrubs**. Flowering shrubs include those that have an ever-flowering nature like *Mussaenda erythrophylla, Hibiscus rosasinensis, Kopsia fruiticosa, Tecoma stans* and *Cassia biflora*. Foliage shrubs having colourful foliage include Crotons, *Cordyline rubra, Poinsettia pulcherrima* and Eranthemums and those having variegated foliage include *Graptophyllum pictum, Acalypha godseffiana,* Dracenas and variegated cultivars of *Pleomele, Aralia* and *Duranta.*

CLASSIFICATION BASED ON LIGHT REQUIREMENT

Based on their adaptation to shaded and open situations in gardens, shrubs can be classed as **Sun-loving shrubs** and **Shade-loving shrubs**. Sun loving shrubs include majority of the flowering shrubs and a few foliage shrubs like crotons. Shade loving shrubs mostly include foliage shrubs which have variegated leaves and a few shrubs which are adapted to low light intensities like *Jacobinea carnea, Jacobinea coccinea, Pachystachys lutea* and *Beloperone guttata*. Partially shade loving are those shrubs that grow well in indirect sunlight. They include *Pentas lanceolata, Brunfelsia* spp. and the Asystasias.

FLOWERING SHRUBS FOR THE TROPICS

The popular **flowering shrubs** of Indian gardens are listed below:

Acalypha hispida Burm. Euphorbiaceae
Syn. *Acalypha sanderi*

Red-hot Cats' tail, Chenille plant (Eng.), Hatisunr (Beng.)

Plants are bushy shrubs growing up to 2m to 3m in height. The leaves are ovate with broadly toothed margins. The flowering spikes, which emerge opposite leaf axils are long, red in colour and droop down gracefully. Plants grow vigorously and need frequent pruning to maintain them in flower continuously. The spikes are branched and are white in var. *alba*. Propagated by stem cuttings. *Acalypha reptans* var. *pygmaea* called the 'Trailing Chenille plant' or 'Strawberry Fire-tails' is a sub-shrub attaining heights below 30cm. Plants thrive in partial shade and are suited for hanging baskets or for growing as a ground cover.

Adenium obesum Roem. & Schult. Apocynaceae

Desert rose

A hardy thick stemmed plant native to the tropical deserts of Africa and N. America. In its native habitat plants attain a height of up to 3m.The stem is grayish white and becomes swollen at the base. Flowers are borne in terminal clusters from June to September in northern India and almost throughout the year in the South. The bell shaped flowers are coloured light to dark pink with reddish streaks on the corolla tube and along the edge of the petals. Propagated by seeds and layers.

Artabotrys odoratissimus R. Br. Annonaceae

Hari Champa (Hindi), Kanthali Champa (Beng.) Manoranjitham (Sans. / Mal.)

Plants are shade adapted woody climbing shrubs and are vigorous in habit. They are among the fast vanishing species that adorned ancient Indian gardens. Fragrant greenish yellow flowers are produced in leaf axils singly or in few flowered clusters almost throughout the year. The corolla is shed entirely and remains picturesquely strewn around the plants. Plants need to be regularly pruned to keep them in shape and to be grown as shrubs. Vegetative propagation is by layers.

Argyreia speciosa Convolvulaceae
Syn. *Argyreia campanulata*

Elephant creeper

Plants are robust and hardy with thick and sturdy, erect to decumbent light green stems of more than a meter length. The shoots terminate in axillary clusters of large white to pale lilac and mauve flowers. The cordate leaves are dull green in colour. Flowered shoots are to be frequently pruned to prevent overgrowth of branches and to remove spent leafless branches. Plants are suited for background planting against bare walls.

Asclepias currassavica Linn. — Asclepiadaceae

Blood flower, Milkweed, Silkweed.

Plants are hardy, partially shade loving, sparsely branched and attain heights of about 1m. The flowers are orange-red and are held in terminal clusters on long erect shoots. The leaves are narrowly lanceolate.. Propagated by seeds and cuttings. They tend to spread and may attain the stature of weeds unless checked.

Asystasia travancorica Bedd. — Acanthaceae

Chinese Violet, Philippine Violet

Plants are sub-shrubs attaining heights below 1m. They are adapted to partially shaded situations. The mauve-purple flowers are bell shaped and are produced in few flowered clusters. A*systasia chelonoides* Nees. has dark purple flowers, which are uniformly streaked with white. These are propagated by cuttings and seeds.

Asystasia coromandeliana Nees. — Acanthaceae

Syn. Asystasia gangetica

Coromandel Primrose, Creeping Foxglove, Ganges Primrose

Plants are straggly sub-shrubs that have ovate-cordate leaves. Purple coloured blooms are held on long racemes. Plants are ideal as ground covers and for hanging baskets.

Barleria cristata Linn. — Acanthaceae

Blue-bell, Philippine Violet (Eng.), Thati (Hindi, Beng.), Thate (Tamil, Kann.), Vella/ Neela Kanakambaram (Mal.)

Plants are robust shrubs. The leaves are ovate and dark green. The bell shaped flowers are white in cultivar 'Candida' and mauve in 'Rosea'. White flowers streaked with purple or purple flowers streaked with white are also seen. The flowers are protected by spiny or sharply pointed bracts. Cultivars that set seed are seed propagated. Natural layers from the base of mother plants are also used for propagation. *Barleria gibsonii* Dalz. has ovate-lanceolate leaves. The flowers are pale purple in colour. *Barleria tomentosa* Roth. has rose coloured flowers.

Bauhinia acuminata Linn. Fabaceae

Orchid tree (Eng.), Mandaram (Mal.).

Plants grow to a height of up to 2m and have attractively bilobed leaves. Pure white flowers that are 4 to 5 cm across, produced throughout the year makes this plant one of the most beautiful of garden shrubs grown along the west coast of peninsular India. Propagated by seeds and layers.

Bauhinia galpinii Fabaceae

The Pride of Cape

Plants are spreading shrubs native to tropical Africa, attaining heights of 2 to 3m. Dark vermilion-red flowers are produced terminally almost throughout the year in tropical gardens. The petals are characteristically clawed. Plants are propagated by seeds and are ideal for group planting.

Bauhinia tomentosa Linn. Fabaceae

Bell-bauhinia, St.Thomas Tree (Eng.), Manja mandaram, Swarna mandaram (Mal.).

Plants are attractive with a compact growth habit and attain heights of up to 2m. The bilobed leaves are small and the half open yellow flowers have a drooping habit. Plants are propagated by seeds.

Beloperone amherstiae Nees. Acanthaceae

Red Justicia, Red Shrimp plant

Plants are sub-shrubs adapted to shaded situations. They have branches terminating in erect rust-pink showy bracts, which enclose less showy flowers. Plants are propagated by cuttings.

Beloperone guttata Acanthaceae

Shrimp plant

Plants are shade loving sub-shrubs attaining heights below 1m. The shoots terminate in inflorescences, which consist of brick coloured bracts that curve gracefully down from the upper half. The brick red colouring and curvature of the inflorescence provide a superficial resemblance to shrimps. Propagated by cuttings. *Beloperone oblongata* Lindl. has bracts that are rose-purple colored and *Beloperone lutea* has greenish yellow bracts.

Bixa orellana Linn. Bixaceae

Kumkumapoovu (Mal.)

Plants are arboreal shrubs having large cordate leaves and sturdy branches. They produce clusters of pink flowers terminally on branches. More attractive than the flowers are the clusters of hairy, dark red fruits. They contain seeds, which are coated with a powdery red dye. The edible dye extracted from the seeds called annato was used in the past to colour silk thread and fabric. Propagated by seeds.

Brunfelsia americana Linn. Solanaceae

American Brunfelsia, Lady of the Night

Plants are native to S. America. They are bushy shrubs producing solitary creamy yellow flowers that have a jasmine-like fragrance.

Brunfelsia pauciflora Benth. Solanaceae
Syn. *Brunfelsia calycina*

Yesterday-Today-Tomorrow plants

These are bushy shrubs, which produce small clusters of scented violet flowers, which change into lilac colour and then to white before starting to fade and dry up. A number of species are grown in tropical gardens and include *Brunfelsia grandiflora* which is large flowered, *Brunfelsia hopeana* Benth. in which flowers have a yellow center and *B. latifolia* Benth. in which they have a white eye. Propagated by layers.

Caesalpinia pulcherrima Sw. Fabaceae
Syn. *Poinciana pulcherrima*

Barbados Pride / Peacock flower / Flamboyant flower (Eng.) Krishan Charan (Hindi), Krishnachura (Beng.), Rajamalli (Tam. /Mal.).

Plants are robust with a dense cover of pinnate leaves. The bright orange flowers are produced in terminal corymbs. They are yellow in cultivar 'Flava' and rosy-red in cultivar 'Rosea'. Plants need drastic pruning to reduce the height of the canopy and to induce flowering in branches at lower heights. Propagated by seeds.

Calliandra Fabaceae

Powder puff / Tassel Flower

Plants are tall shrubs with pinnate leaves on spreading branches. The flowers have showy radiating stamens, which are held like a powder puff. The flowers are light pink in *Calliandra brevipes* Benth. and bright red in *C. haematocephala* Hassk. and pure white in *C. portoricensis* Benth. *C. emarginata* Benth. is a sub-shrub having small leaves and dark red flowers. They are ideal for pot growing. All the species can be propagated by air layering

Carissa carandas Linn. Apocynaceae

Christ's thorn, Caronda (Eng.), Karamcha (Beng.), Koronda, Karaunda (Hindi), Karavanda, Boronda (Mar.), Oka, Vaka (Tel.), Heggarichige (Kann.) Poo kara (Mal.).

A vigorous plant, which has a dense cover of shiny bright green leaves on thorny stems. The flowers are white with dark pink corolla tubes. More attractive than the flowers are the fruits, which turn from white to light pink and to dark pink before drying and shedding. Propagated by cuttings which root readily under mist.

Cassia alata Linn. Fabaceae
Syn. *Senna alata*

Aavaram poo (Tam.), Golden Candlesticks, Golden Brush plant (Eng.)

Plants are quick growing and erect. The pinnate leaves carry large leaflets. The shoots terminate in large well-formed racemes of golden yellow flowers. Propagated by seeds. Plants are native to tropical America.

Cassia biflora Linn. Caesalpinaceae

Twin-flowered Cassia

Plants are profusely flowering shrubs native to Central and South America. The leaves are small, pinnate and are closely arranged on arching branches. Clusters of golden yellow flowers are produced in leaf axils all along the branches. Propagated by air layering.

Cestrum nocturnum Linn. Solanaceae

Raat-ki-Raani (Hin.), Night blooming Jasmine, Night Cestrum (Eng.)

Plants attain heights of 2 to 2.5m and bear alternate leaves. The fragrant flowers are borne in terminal and axillary clusters. The tubular and five-lobed corolla is greenish white. The flowers open at night. *Cestrum diurnum* Linn. called 'Din-ka-Raja' has glabrous shiny leaves and fragrant white flowers that open during daytime. Both the species are native to West Indies and have been introduced into Indian gardens.

Clerodendrum fragrans Willd. Verbenaceae

Forest jasmine (Eng.), Jendu Peruvalam (Mal.)

Plants are adapted to partially shaded situations. They have dark green, cordate leaves held on long stalks around the main stem. From the leaf axils close to the main stem arise dense clusters of very fragrant white flowers. The calyx and flower pedicels are tinged maroon-red. It is a favoured shrub of traditional Indian gardens.

Clerodendrum inerme Gaertn. Verbenaceae

Forest jasmine (Eng.), Peruvalam (Mal.)

A sparsely branched shrub attaining heights of up to 2m, these are found naturalized throughout the humid tropics of peninsular India. Shoots bear pyramidal paniculate inflorescences carrying lightly scented white flowers.

Clerodendrum macrosiphon Hook. Verbenaceae
Syn. *Clerodendrum inscisum*

Musical Note flower, Witches tongue

Plants are compact shrubs attaining heights of about 1m and are ever flowering. The leaves are light green and are easily bleached to attain a yellowish tinge in full sunlight. Plants thrive in partial shade. Clusters of white flowers are produced throughout the year. The flowers have purplish red stamens. Plants are suited for making flowering hedges.

Clerodendrum nutans Wall. Verbenaceae
Syn. *Clerodendron wallichii*

Plants grow to a height of 1.5m and have four-angled branches. The leaves are glabrous and oblong-lanceolate with slightly wavy margins. The flowers are white, non fragrant and are borne in gracefully drooping panicles.

Clerondendrum siphonanthus (R. Br.) Ait. Verbenaceae
Syn. *Clerondendron indicum*

Turk's turban, Tube flower

Plants are erect shrubs attaining heights of up to 2.5m. The leaves are lanceolate. The branches carry clusters of tubular flowers with coiled tips. Plants thrive in partial shade too. Propagated by cuttings.

Clerodendrum squamatum Vahl. Verbenaceae

Clerodendron kaempferi Fischer ex Morr.

Krishna Kireedam (Mal.).

A sparsely branched shrub attaining heights of up to 2m, these are found naturalized throughout the humid tropics of peninsular India in shaded situations. Shoots bear terminal paniculate inflorescences containing bright red flowers. The pyramidal inflorescences have decorative value.

Crossandra infundibuliformis (Linn.) Nees Acanthaceae
Syn.*Crossandra undulaefolia* Salisb.

Fire-cracker plant (Eng.) Kanakambaram (Tam. /Mal.).

Plants are sub-shrubs, which are adapted to grow in partial shade and in the open. They are commercially cultivated for their flowers that have decorative value. Flowers are borne in spikes enclosed within bracts. Their colours vary from yellow to dark orange. Delhi Crossandra is a dark orange-coloured variety. Co.1 is a yellowish-orange variety evolved at the Tamil Nadu Agricultural University and Soundarya is a pink coloured variety. The dried pods break open with an audible crackle. Propagated by seeds and herbaceous cuttings. *Crossandra nilotica* Oliver, the Red Fire-cracker plant has brick-red flowers.

Cuphea hyssopifolia H.B.&K. Lythraceae

False Heather

Dwarf hardy shrubs having thin erect branches and closely spaced small leaves. Plants are also grown in shade. The flowers are mauve in *Cuphea aquipetala* Cav. (=*Cuphea miniata* Brongn.) and are white in *Cuphea alba*. *Cuphea ignea* is the Fire Bush that has tubular red flowers tipped with yellow and is adapted to shaded situations.

Dianthera nodosa Acanthaceae

Pink Justicia

Plants have light green leaves held on arching branches. They are adapted to shade and can be maintained ever flowering by adequate pruning. Clusters of tubular pink flowers arise from the nodes.

Dombeya spectabilis Boj. Sterculiaceae

Dombeya, Snow ball flower

A large shrub with a canopy of cordate, lobed leaves which are dull green in colour. Large bunches of white to pale pink flowers are produced from November to February. Excessive vegetative growth is to be pruned after flowering. Plants are native to tropical Africa. Propagated by air layering.

Duranta plumieri Jacq. Verbenaceae
Syn. *Duranta repens* Linn..

Golden Dewdrop, Sky Flower

Plants are compact, bushy and attain heights of up to 2m or more. The branches bear blue flowers in elongated clusters. The fruit bunches droop downwards gracefully making plants particularly attractive in the fruiting season. *Duranta erecta* 'Goldiana' is extensively grown at present as a foliage hedge. It has yellowish green young leaves, which eventually turn dark green. The flowers are blue. They are white in var. *alba*. Variants with yellow blotches on the leaves or a white edging along the margins are included in var. *vareigata*. Plants are propagated by terminal herbaceous cuttings.

Euphorbia mili Euphorbiaceae
Syn. *Euphorbia splendens*

Crown of Thorns, Christ's Thorns

Plants are succulent shrubs that attain heights of up to 1.5m or more. The shoots are brown and spiny and carry ovate leaves that are clustered together. The flowers are small and two floral bracts that subtend them are showy and are coloured white or pink to red. The cultivar 'Minibell' has smaller leaves and flowers and is called 'Miniature Crown of Thorns'.

Euphorbia x *lomi* Euphorbiaceae

Flowering Euphorbias

These are large flowered hybrids of *Euphorbia lophogona* and *Euphorbia mili* native to Madagascar. They are popular throughout the tropics as garden plants. The floral bracts are white or pink to red, orange and peach, attractively mixed with green.

Galphimia gracilis Bartl. Malpighiaceae

Golden Shower Bush, Golden Thyrallis, Rain of Gold, Slender Golden Shower

Plants are sub-shrubs having red tinged thin branches and small ovate leaves. The flowers are small and golden yellow in colour and are produced in profusion terminally. Close planting is to be done to prevent lodging of plants in heavy rains. Propagated by seeds. Plants are native to Mexico. Ideal for flowering hedges, they are to be left untrimmed to enable profuse flowering.

Gardenia jasminoides Rubiaceae
Syn. *Gardenia florida* Linn. / *Gardenia radicans* Thun.

Cape jasmine (Eng.), Gandharaj (Beng.), Gandharajan (Mal.).

Plants are favoured scented flowering shrubs of the humid tropical and sub tropical belt of India. They are arboreal in stature, with a dense cover of shiny green leaves. The flowers are pure white, fading to an off-white colour and are very fragrant. Varieties having fully double flowers (**flor-pleno**) and those having a compact flower form with slowly opening imbricate petals are also grown in gardens. *Gardenia indica* has smaller and less fragrant flowers.

Hamelia patens Jacq. Rubiaceae
Syn. *Hamelia coccinea* Sw.

Humming Bird bush, Scarlet bush

Plants are compact ever-flowering arboreal shrubs. The leaves are dark green. The branches are short and sturdy and they carry terminal clusters of tubular orange-red flowers. They can be trimmed and maintained as flowering hedges. Propagated by cuttings.

Hibiscus mutabilis Malvaceae

Changeable rose (Eng.), Gulab-i-Ajaib, Thalkamal (Hindi), Sthalpadma (Ori.) Thalopodmo (Beng.), Bettada tavare (Kann.)

A tall growing open shrub with limited branching. The lobed leaves are dull green in colour. The large white fully double flowers are produced terminally on the axils of the branches. They are white at the time of opening early in the morning and when day temperature rises, they turn pink and by the afternoon, become deep rose pink and start fading by the evening of the same day. Easily propagated by cuttings. Adapted to the humid tropical belt of India.

Hibiscus rosasinensis Linn. Malvaceae

China rose, Shoe flower (Eng.), Gurhal, Jasut (Hindi), Jaba kusum (Beng.), Jaba (Assa.), Jassvandi (Mar.), Mondaro (Oriya), Dasavala (Kann.), Sembaruthi (Tam.), Chembarathi (Mal.).

Popularly grown bushy shrub of Indian gardens, these attain heights of 2 to 2.5m. The branches are strong and woody and the leaves are broadly ovate with serrate margins. The flowers maybe single to fully double. In singles, the five petals are characteristically reflexed with a prominent arching staminal column, which bears numerous stamens toward the tip below the multifid stigma. A large number of hybrid varieties and bud sports are grown with a range of flower colours and with ruffled or wavy petals. These include varieties introduced from Hawaii and hybrids released by the Indian Institute of Horticultural Research such as 'Arunodaya' (single- nasturtium-orange with mauve-pink along the veins), 'Basant' (single, sulphur-yellow), 'Be-Nazeer' (single, neyron-rose to capsicum-red towards the upper half), 'Geetanjali' (single, dark red to turkey-red towards the upper half), 'Nartaki' (double-cardinal red to marigold-orange towards the tips), 'Pakeezah' (single-claret rose to carmine-red towards the tips). A variety with variegated leaves having white and green patches having single dark red flowers is also grown in gardens. Plants are propagated by cuttings, layers and budding.

Hibiscus schizopetalus Hook. Malvaceae

Jhumko jaba (Beng.), Thooku Chembarathi (Mal.).

Plants resemble *H. rosasinensis* in vegetative characters. The single red flowers have petals with cut margins, which are completely reflexed. As a result, the entire flower along with the staminal column droops gracefully.

Hibiscus syriacus Linn. Malvaceae

Mauve, Pink and White hibiscus

Plants are less sturdy than *Hibiscus rosasinensis* in habit. Upright branches that are sparsely leaved, bear shoots angularly. The leaves are small and have deeply serrate margins. Seed set is observed. However, plants are propagated by cuttings.

Hibiscus tiliaceous Malvaceae

Tree Hibiscus, Beach Hibiscus

Plants attain heights of up to 4m. They are bushy having dull green heart shaped leaves with prominent basal lobes and faintly serrate margins. The lamina has a characteristically rough texture on the underside, the main veins are red tinged and the yound leaves have prominent ovate stipules. Plants are grown as tall hedges alongside waterways and in sea-coasts. They are often used as a source of green manure. Flowers are produced in terminal leaf axils. They are cupped and golden yellow with maroon-red centers. They open at night and during the next day darken to a light orange colour and start furling and fading by evening. The cultivar "Variegata" is gron ornamentally and has attractive pink and reddish green new leaves that turn creamy white and green with age.

Holmskioldea sanguinea Retz. Verbenaceae

Cup and saucer flower, Parasol Flower

Plants are bushy and have long arching branches, which carry small ovate leaves. The small reddish orange or yellow flowers have a saucer-shaped calyx, with a centrally placed cupped corolla tube. Flowers are produced all along the leaf axils and branches. Plants flower profusely in sub-tropical to warm-temperate situations. They are also pruned and grown as flowering hedges. Propagated by air layering. The flowers are pale orange in *H. sanguinea* var. *accuminata* and red in *H. coccinea*

Hydrangea hortensia Siebold. Saxifragaceae
Syn. *Hydrangea hortensis* Sm.

Hydrangea

Plants are sturdy and carry attractively serrate bright green leaves. Inflorescences are in terminal globose bunches and carry small pink to blue flowers. The pH of the growing medium can influence the pigmentation of petals. In alkaline pH, the petals are cloured pink and in acidic pH, they are mauve-blue. Plants are heavy feeders. In the humid tropical belt of South India, plants thrive and flower well under partially shaded situations. The flowered shoots are to be cut back periodically. *Hydrangea arborescens* Linn. (=*Hydrangea paniculata* Rafin.) called the Japanese Hydrangea is a more vigorous and cold hardy species adapted to higher elevations. The flowers are white, attaining tints of mauve and blue. Plants are propagated by stem cuttings.

Ixora acuminata Roxb. Rubiaceae
Syn. *Ixora parviflora* Vahl.

White Ixora

Plants are compact arboreal shrubs attaining heights of up to 3m. The branches are short, sturdy and bear dark green leaves. Large globose heads of faintly scented flowers are borne terminally on the branches. They are traditionally grown in home gardens of the western coast of peninsular India.

Ixora chinensis Lam. Rubiaceae

Dwarf Ixora

Plants have a compact growth habit and attain a height of 1.5 m. The branches are short, upright and have closely spaced small leaves. Axillary inflorescences are produced in profusion and carry small red, pink or yellow flowers. Propagated by herbaceous cuttings.

Ixora coccinea *Linn.* Rubiaceae
Syn. *Ixora arborea* Lodd.

Rukmini (Hindi), Thetti (Mal.)

Plants are compact attaining heights of upto 1.5m. Propagated by seeds and cuttings. The leaves are ovate and dark green. A popular garden shrub of India.

Ixora singaporensis Rubiaceae

Asoka-thetti (Mal.)

Plants are tall and vigorous with long and broadly ovate leaves, which are bright green in colour. The branches carry large globose terminal inflorescences of bright red flowers. A handsome garden shrub, which is popular in the tropical regions. Propagated by air layering.

Jacobinea carnea Acanthaceae
Syn. *Justicia carnea*

Pink Jacobinea, Pink Justicia

Plants are dwarf in habit and adapted to shade. The branches are short and leaves are broadly ovate and pale to bright green. Almost all the branches terminate in inflorescences carrying tubular pink flowers, which are enclosed in green bracts. Propagated by herbaceous cuttings.

Jacobinea coccinea Acanthaceae
Syn. *Justicia coccinea*

Red Jacobinea Red Justicia

Plants attain heights of up to 3m or more and are adapted to shade. The leaves are dark green. The flowers are tubular, bright red and are produced terminally. Dark green bracts subtend flowers.

Jasminum auriculatum Vahl. Oleaceae

Joohi (Hindi), Mullai (Tam.), Arimulla (Mal.)

Native to the subcontinent, plants are shrubby and tend to climb when supported. They have trifoliate ovate leaves with slightly wavy margins. Flowers are borne in few flowered cymes. They are have thin corolla tubes and narrow petal lobes and are less scented than several other species.

Jasminum grandiflorum Linn. Oleaceae

Catalonian, Italian, Spanish Jasmine (Eng.), Chameli, Kuth Bela, Kund, Jati, Jalee (Hindi), Malati (Sans.), Pitchi (Tam. / Mal.)

Indigenous to India, plants are large and bushy with long arching branches which carry pinnate leaves having five to seven leaflets. The white flowers have their outer surface tinged pink. They are exquisitely fragrant and are produced in three to eight flowered cymes. Leafless branches that lie on the soil surface root easily. Plants are very popular garden shrubs throughout India.

Jasminum multiflorum Roth. Oleaceae
Syn. *Jasminum pubescens* Willd.

Dela (Hindi), Kattu mulla (Mal.)

Plants are vigorous with arching branches that terminate in large bunches of pure white flowers which are faintly fragrant. The young branches are downy.

Jasminum sambac (Soland.)Ait. Oleaceae

Arabian Jasmine, Tuscan Jasmine (Eng.) Bel-Bela, Bela, Motia (Hindi), Malli, Gundumalli (Tam.), Mulla, Kudamulla (Mal.)

Plants are evergreen sub-shrubs having glabrous ovate-elliptic leaves with entire margins. Flowers are borne in few-flowered cymes. They may be single, semi-double or fully double and are very fragrant. Plants need to be pruned regularly as flowering occurs in newly emerged branches only. The cultivar 'Mogra' has creamy white fully double large sized flowers.

Jatropha hastata Jacq. Euphorbiaceae
Syn. *Jatropha acuminata* Desr. / *Jatropha integerrima / Jatropha panduraefolia* Andr.

Peregrina, Peregrine

Plants are tall growing with open branches carrying ovate-lanceolate leaves. They are native to the West Indies. The flowers are small and bright red in colour and are produced in compact clusters. Propagated by cuttings and layers.

Jatropha multifida Linn. Euphorbiaceae

Physic nut (Eng.), Bherenda (Beng.).

Plants are tall growing and have deeply cut lobed leaves with long petioles. The flowers are bright red in colour. Propagated by seeds and cuttings. Plants are native to S.America.

Jatropha podagrica Hook. Euphorbiaceae

Gout plant, Bottle plant, Guatemala Rhubarb, Physic nut, Tartago (Eng.)

A stout stemmed hardy plant with large palmately lobed leaves, which are held on long petioles. The main stem becomes bulged at the base with ageing. The small flowers are bright red in colour. They are borne in clusters on long stalks. Plants are popularly grown in rock gardens.

Justicia coccinea Aubl. Acanthaceae
Syn. *Pachystachys coccinea, Jacobinea coccinea* Hiern.

Cardinal's Guard

Plants are tall growing and adapted to shades. Leaves are long, broadly ovate and dark green in colour. Terminal inflorescences carry scarlet flowers that are enclosed in dark green bracts. In shaded situations where plants are massed the flowers are spectacular. Propagated by cuttings.

Justicia gendarussa Burm. Acanthaceae

Black Adusa, Gandarusa, Malabar nut, Violet Justicia

Plants are short in stature, growing up to a height of 1m. They are hardy with linear narrowly lanceolate leaves, which become violet tinged when grown in the open. The flowers are white with faint purple spots. Popular flowering hedges of the past, they are propagated by cuttings.

Kopsia fruticosa A. DC. Apocynaceae

Pink Kopsia, Shrub Vinca

Plants are tall growing with uniformly thin-stemmed branches, having long internodes. The leaves are ovate-lanceolate. The pink flowers are produced in showy clusters throughout the year. Plants can be pruned and maintained at desirable heights, in the open and in partial shade. Propagated by cuttings.

Lagestroemia indica Linn. Lythraceae

Crape myrtle, Queen flower, Pride of India (Eng.), Gul-i-Phanos (Hindi)

Plants are tall with thin upright shoots and sparse branching. The leaves are small and ovate. Lilac, pink or white flowers with crinkled fringed and wavy petals are produced in bunches terminally on the shoots during the summer months. These are often designated respectively as var. *purpurea*, var. *rosea* and var. *alba*. Propagated by cuttings and layers.

Lantana camara Linn. Verbenaceae

Wild Sage (Eng), Kuri (Hindi), Poochedi (Mal.)

Plants are bushy with spiny arching branches. The short secondaries and tertiaries produce pink to yellow or orange-red flowers throughout the year. Plants are to be pruned to restrict the number of shoots and to induce short branches that will flower readily. The species sets seeds profusely. This has led to its spread like a noxious weed in many tropical countries. However, improved dwarf varieties grown in gardens fail to set seed and need to be propagated by cuttings. *Lantana crocea* Jacq. resembles *L. camara* in stature and has flower heads carrying bright red flowers towards the periphery and yellow flowers towards the center. *Lantana camara* var. *depressa* is short in stature and produces golden yellow flowers in profusion, which makes it a garden favourite. *Lantana sellowiana* Link. & Otto, the trailing lantana, with smaller leaves and smaller clusters of lilac or white flowers has a decumbent habit and is grown as a creeper or trained over supports.

Malpighia coccigera Linn. Malpighiaceae

Indian Holly

Plants are hardy with strong and thin upright stems. The small shiny leaves have sharply pointed margins which make the leaves seem spiny. The small flowers are pale pink in colour and are produced in profusion, covering the arching branches entirely. Plants are amenable to pruning and are frequently maintained as cattle-proof boundary hedges.

Malvaviscus arboreus Cav. Malvaceae

Mulaku Chembaruthi (Mal.)

Plants are hardy and bushy in stature. They produce a profusion of bright red or pale pink flowers throughout the year. The flowers remain unopened like buds. Plants are frequently grown as flowering hedges. Propagated by cuttings.

Melastoma malabathricum Linn. Melastomaceae

Malabar Black Mouth

Plants are sturdy, attaining heights of up to 1m in full sunlight. They are found growing along waysides near low-lying areas. The leaves are ovate, with prominent and depressed parallel veins. The flowers are mauve-purple with five showy petals.

Murraya exotica Linn. Rutaceae
Syn. *Murraya paniculata* Jacq.

Orange Jasmine, St. Sebastian's flower.

Plants are arboreal shrubs with attractive shiny foliage. The branches are thin and amenable to pruning. Plants are grown as hedges and for making topiary. Propagated by cuttings.

Mussaenda erythrophylla Schum. & Thonn. Rubiaceae

Mussanda

Plants are ever-flowering shrubs and are popularly grown as focal point plants in small gardens and shrubberies. The cultivated varieties include 'Red' in which the sepals are scarlet in colour, 'Rosea' in which they are bright pink and 'Queen Sirikit' in which the sepals are pale pink with a fine red line along the edge. *Mussaenda glabra* Hook. & Vahl. (Syn. *M. frondosa* Blanco / *Mussaenda grandiflora* Benth. / *Mussaenda pubescens* (Dryand) Ait.) has yellow flowers and a single white sepal and *Mussaenda luteola* Delile., which is short statured, has a single yellow sepal, while *Mussaenda philippica* A. Rich. has clusters of large velvety white sepals.

Neerium oleander Apocynaceae

Oleander, Rose Bay (Eng.), Arali (Mal./Tam.)

Plants are tall and elegant attaining heights of more than 4m, with narrowly-lanecolate and leathery dark green leaves. Large bunches of flowers are produced terminally throughout the year. The flowers are single to double and white to light pink and dark rosy pink in colour. Propagated by cuttings in the rainy season. Air layering is done on the improved varieties which have large bunches of single flowers.

Neerium odorum (Soland.) Ait. Apocynaceae

Oleander

Plants are shorter in stature when compared to *Neerium oleander*. The flowers are single and coloured white to dark pink. They are produced in large clusters of 50 to 70.

Ochna kirki Oliver Ochnaceae

Mickey Mouse flower

Plants attain heights of up to 3m and have a compact growth habit. The branches are short and sturdy and carry shiny leaves with finely serrate margins. The flowers are yellow and the round green fruits develop on an enlarged torus, which is bright red in colour. The fruits eventually turn black. Birds relish them.

Ochna squarrosa L. has a similar stature and yellow flowers.

Odontonema cuspidatum Acanthaceae

Cardinal's Guard, Mottled toothedthread, Odontonema

Plants are adapted to partial shade. The inflorescence stalk is long and erect, bearing tubular red flowers.

Otacanthus caeruleus Plantaginaceae

Butterfly flower, Brazilian Snapdragon, Caribbean Blue

These are partially shade loving sub shrubs that attain heights of up to 75cm. The leaves are ovate with crenulated margins. The flowers are a brilliant-blue coloured with two rounded to slightly wavy wing-like lobes. Plants are ideal for pots and for bedding.

Pachystachys lutea Nees. Acanthaceae

Candlestick plant

Plants are ever-flowering sub-shrubs in which all the shoots terminate in inflorescences, which have showy golden yellow bracts. The flowers that emerge from them are less showy and white in colour. Plants are well adapted to shady situations and are grown in pots or close planted in beds. An ideal flowering hedge for shaded situations.

Pentas lanceolata Benth. Rubiaceae

Pentas

Plants are sub-shrubs adapted to partial shade. The shoots are thin and hairy and become hollow with ageing. Leaves are ovate with well-marked veins and are also finely hirsute. The flowers are produced in clusters. Inflorescences are corymbose cymes. Flowers are white in var. *alba*, pink in var. *quartiniana* and carmine-red in var. *kermesiana*. Plants are native to tropical Africa and are propagated by herbaceous cuttings.

Plumbago capensis Thunb. Plumbaginaceae

Leadwort (Eng), Neela Koduveli (Mal.)

Plants are evergreen and adapted to partially shaded situations. Pale blue colored flowers are produced in terminal clusters almost through out the year.

Plumbago rosea Linn. Plumbaginaceae

Chitra, Chitra mulla

Plants are sub-shrubs with elongated arching shoots and small leaves. The branches carry clusters of rosy-red flowers. Plants are adapted to partial shade and can be pruned to form hedges. *Plumbago rosea var. coccinea* (chuvanna koduveli) has larger scarlet red flowers and is adapted to higher elevations.

Plumbago zeylanica Linn. Plumbaginaceae
Syn. *Plumbago auriculata* Blume.

Vella koduveli (Mal.)

Plants are short statured and have angular stems. The flowers are white and are produced in small clusters throughout the year.

Quassia amara Linn. Simaroubaceae

Amargo, Bitter Quassia, Surinam Quassia

Plants are sturdy, growing up to a height of 4m, with open branches carrying elongated pinnately compound leaves. The flowers are bright red in colour and are held in drooping clusters. Propagated by seeds, cuttings and layers. Plants are native to tropical America.

Randia maculata DC. Rubiaceae

Randia, Star flower

Plants are arboreal shrubs with sturdy branches and small ovate leaves. They are amenable to pruning and are hence made into hedges. Their closely spaced leaves make them suitable for topiary too. The flowers are pure white and have a long corolla tube and star-shaped free lobes. Propagated by cuttings and layers.

Reinwardtia indica Dum. Linaceae
Syn. *Reinwardtia trigyna* Dalz. & *Gibs.*

Yellow Flax

Plants are sub-shrubs with ovate leaves. They produce yellow flowers almost throughout the year in peninsular India. Flowering is from November to February in northern India. They are suited for growing in mixed borders. Propagated by cuttings.

Rondeletia odorata Jacq. Rubiaceae
Syn. Rondeletia speciosa Lodd

Fragrant Panama rose

Plants are bushy shrubs resembling *Ixora coccinia* in habit. They are adapted to partial shade and open situations. The flowers are orange-red with rounded petals and a yellow eye at the center. Propagated by layering. *Rondeletia odorata* var. *breviflora* has smaller flowers.

Rosa spp. Rosaceae

Shrub roses, Bush roses

Rose plants that attain the stature of shrubs or bushes are more preferentially grown and displayed in tropical gardens than climbers ramblers or miniatures. Bushy plants with evergreen foliage are used for bedding while large shrub roses are displayed as specimen plants in lawns or as focal point plants in small gardens. They include Floribundas and Hybrid Teas. Apart from these, fragrant cultivars like the 'Edouard Rose'- *Rosa bourboniana* and the pink-red flowered *Rosa centifolia* are favourites of suburban gardens in South India.

Ruellia L. Acanthaceae
Syn. *Aphragmia* Nees.

Ruellia

This is a genus of sub-shrubs, adapted to partially shaded situations. Branching is sparse and the leaves are narrowly lanceolate. In *Ruellia formosa* the flowers are red in colour. Propagated by seeds. *Ruellia lorentziana* Griseb. has lilac coloured flowers. *Ruellia macarantha* Mart ex Nees.. has light bluish-purple flowers. It is a hardy dwarf annual shrub, which sets seed profusely. *Ruellia tweediana* Griseb. is a perennial and an ideal edge plant having leaves that are held in a rosette. The flowers are pink in colour.

Russelia juncea Zucc. Scorphulariaceae
Syn. *R. equisetiformis* Schlet.& Cham.

Coral Bush, Wedding Bells

Plants are bushy shrubs with thin wiry green whorled branches, which are arching in habit. The leaves are much reduced and appear like bracts. The branches bear loose clusters of tubular coral-red flowers. Suited for planting in rock gardens. Propagated by division of shoots having roots from the base of the plants. *Russelia rotundifolia* Av. attains a height of 1.5m and has round, cordate leaves. The flowers are red in colour and are borne terminally on upright shoots.

Schaueria flavicoma Acathaceae

Brush flower

Plants are sub shrubs adapted to partial shade. The flowers are tubular, white and are borne on rounded inflorescences covered by light green brush like bracts.

Serissa fetida Lam. Rubiaceae
Syn. *Serissa japonica* Thunb..

Japanese Serissa, Serissa, Yellow-rim flower

Plants are bushy and dwarf in stature. Flowers are small white and star-like. They are produced in rounded clusters almost throughout the year. Propagated by cuttings

Strobilanthus Blume. Acanthaceae

Kurinji

This is a genus of hardy sub-shrubs adapted to partially shaded regions from the coastal to the mountainous tracts of the subcontinent. The leaves are ovate and rough textured. The stem has swollen joints. Flowers emerge enclosed in showy papery bracts. They are produced terminally from October-November onwards in the photosensitive species. A number of species are grown as ornamentals. They include *Strobilanthus anisophyllus* T. Anders. having lavender blue flowers, *Strobilanthus colouratus* T. Anders. having petals which are dark green above and purplish below, *Strobilanthus dyerianus* in which the leaves are blotched purple and the inflorescences are pale lilac in colour, *Strobilanthus glomeratus* T. Anders. having mauve bracts and bluish purple flowers, *Strobilanthus isophyllus* having blue or white flowers and *Strobilanthus scaber* Nees. having yellow flowers. *Phlebophyllum kunthianum* Nees. the allied genus is the famed 'Neela Kurinji' endemic to the Western Ghats of S. India, which extensively carpets the native hills with mauve-blue flowers once in twelve years.

Tabernaemontana coronaria Willd. Apocynaceae

Cape flower, Adams Apple, East Indian Rose Bay, Nero's crown (Eng.) Chandani (Hindi), Nandiyarvattom (Mal.).

Plants are hardy and bushy in nature with attractive glossy green foliage. Pure white semi-double flowers are produced profusely in few flowered clusters. The cultivar 'Flor-Pleno' is fully double and is less profuse in flowering. The plants are also taller, attaining heights of up to 2m or more and have larger leaves and branches. These are very popular flowering shrubs of home gardens along the western coast of India. They can tolerate frequent severe pruning and are propagated by cuttings and root suckers.

Tabernaemontana indica Willd. ex. Steud. Apocynaceae
Syn. *Tabernaemontana divaricata*

Nandiyarvattom (Mal.), Eastern Blue Star, Pinwheel flower (Eng.)

Plants are tall arboreal shrubs with shiny leaves. The flowers are single and smaller than those of *Tabernaemontana coronaria*. The petals are slightly twisted to one side. The flowers are produced in large clusters and when shed they carpet the bases of plants. Smaller bushy varieties that have profuse flowering habit and dwarf types that are adapted to shade - cv. 'Nana' with small dark green leaves and flowers that are mostly produced singly - are also grown. 'Silver Rain' is a shot statured cultivar having ash-green leaves.

Tecoma stans Bignoniaceae

Yellow bells, Yellow Elder (Eng.), Gaudi-Chaudi (Hindi)

Plants are tall shrubs with upright branches and a dense foliage cover. The leaves are bright green and have serrate margins. The bell-shaped golden yellow flowers are produced in clusters throughout the year in the western coast. Plants require pruning to restrict their height and spread. Propagated by cuttings.

Tecomaria capensis Spach. Bignoniaceae
Syn. *Tecoma capensis* Lindl.

Cape Honeysuckle

Plants are bushy shrubs with dark green leaves that completely cover branches. The bright orange flowers are produced in clusters almost throughout the year. Plants can be pruned to give the canopy a suitable shape.

Thevetia nereifolia Juss. ex Steud. Apocynaceae
Syn. *Thevetia peruviana*

Yellow oleander (Eng.), Pila Kaner (Hindi, Punj.), Kal ke phool,

Haldi ka rabi (Bengali), Manja Arali (Mal.).

A tall growing shrub having sparse foliage cover made up of narrow leaves which are thin and light green in colour. The flowers are bell-shaped, produced terminally on the shoots. The fruits are characteristically smooth angled. They are poisonous. Plants are usually used for background planting or as tall-untrimmed hedges.

Thunbergia erecta T.Anders. Acanthaceae

Bush Thunbergia, Bushy Clock-vine

Plants are bushy shrubs with erect or arching shoots, amenable to pruning. The leaves are dull green, small and ovate. The flowers are have a pale yellow corolla tube, which ends in bright purple coloured lobes. Plants are used to make hardy flowering hedges. The flowers are white in var. *alba*. Propagated by cuttings.

Tibouchina elegans Cogn. Melastomaceae
Syn. *Tibouchina urvilleana*

Lasiandra, Princess flower, Purple Glory bush

Plants are shrubs suited to sub-tropical and temperate situations. They grow up to a height of 2m. The leaves are finely hirsute and appear velvet like. The flowers are about 3cms in diameter and possess a dark purple colour, which is rare in tropical shrubs. *Tibouchina semidecandra* is another floriferous species grown in gardens. Flowering starts from October-November onwards and continues throughout the summer. Plants are native to Brazil. They are not very common in gardens of the plains and have recently been introduced into the west coast of southern India.

Tithonia diversifolia A. Gray Asteraceae

Mexican Sunflower, Perennial Sunflower

Plants attain heights of up to 5m. They are native to Central America. Leaves are glabrous and 3 to 5 lobed. Flowers are yellow, sunflower-like and are produced in profusion on pinching new shoots periodically.

Trachelospermum jasminoides Lem. Apocynaceae

Confederate Jasmine, Star Jasmine, Tree Jessamine

Plants are evergreen with elongated thin branches. The elliptic lanceolate leaves have are glossy and dark green in colour and acute tips. Flowers are produced terminally in loose clusters. They are five lobed with white petals that have narrow bases and appear slightly twisted in a clock-wise direction. The corolla tube is creamy yellow in colour.

Turnera ulmifolia Linn. Turneraceae

West Indian holly, Sage rose.

Plants are hardy dwarf statured, almost weed-like. The leaves are dull green. The flowers are solitary, pale yellow in colour with a maroon red center. Propagated by seeds.

Vinca rosea Linn. Apocynaceae
Syn. *Catharanthus roseus* G. Don.

Periwinkle (Eng.), Sadabahar (Hindi), Nithyakalyani, Savanari (Mal.)

Plants are hardy sub-shrubs that have glossy green leaves with prominent veins. The flowers are white, mauve-purple or dark pink in colour and have a disagreeable odour. Propagated by seeds and cuttings.

FLOWERING SHRUBS FOR THE HILL ZONES

Evergreen or deciduous shrubs, which flower and thrive in the hilly tracts of India are listed below. Many of them are early introductions into the country and have adapted to grow in a range of ecological situations. However, profuse flowering in response to low temperatures of the hill zones and successful fruit set and germination of seeds observed there are not seen in areas where tropical temperatures and climatic conditions prevail.

The plants listed below are generally seen in the Hill stations of India and are displayed in their magnificence from spring to autumn.

Abelia R. Br. Caprifoliaceae

Abelia

Plants are deciduous attaining heights of up to 1.5m. They thrive in partially shaded situations. The flowers are scented, white to pink in colour and have persistent brown sepals. Plants are propagated by cuttings. Many cultivars of the Chinese Abelia *Abelia chinensis* R. Br., the Mexican Abelia *Abelia floribunda* Decne. and *Abelia schumannii* are grown in gardens.

Abutilon Tourn. ex Adans. Malvaceae

Chinese Bell flower, Chinese lantern, Flowering Maple

Plants thrive in partially shaded situations attaining heights of up to 1.5 m or more. Flowers resemble those of Hibiscus. A number of ornamental species are grown.

Callicarpa cana Linn. Verbenaceae

American Beautyberry, Callicarpa, French Mullberry

Plants are deciduous having a spreading habit and are found to attain heights of over 2.5m. The flowers are mauve to lilac in colour and are axillary in position. *Callicarpa americana* with blue flowers and *Callicarpa japonica* with white flowers are also grown.

Calluna vulgaris Ericaceae

Calluna, Heather, Scotch Heather

Plants are spreading evergreens attaining heights of up to 2m. The foliage attains autumn-like tints seasonally. The flowers are pink to mauve in colour and are borne in erect racemes. Flowers are white in *Calluna vulgaris* var. *alba*.

Camellia Linn. Ternstroemiaceae

Camellia

Plants are evergreen arboreal shrubs adapted to partially shaded situations. They are spring flowering. The flowers are rounded, pink or white in colour and are semi-double to double. They are suited for partially shaded situations and are ideal hedges and specimen plants.

Cestrum L. Solanaceae

Cestrums, Day Jasmines

This is a genus of profusely blooming evergreens attaining heights of up to 3m. *Cestrum aurantiacum* Lindl.has drooping branches with orange-yellow flowers. *Cestrum diurnum* called the Day Queen is erect in habit with scented flowers that are white in colour. *Cestrum elegans* has purplish red flowers, which are followed by purple coloured berries. *Cestrum parqui* has yellow flowers.

Dombeya Cav. Sterculiaceae

Dombeya, Snowball flower

This is a genus of shrubby to arboreal evergreens that attain heights of over 3m. They have a uniformly spreading habit. Leaves are cordate, many lobed and have a coarse surface texture. Flowering takes place in spring or winter. The flowers are white to delicate- pink tinged. A number of species are grown in their native habitat in tropical Africa. Important species grown in India include *Dombeya aliporensis* and *Dombeya lancasteri* with dark pink flowers. Plants are suited for specimen planting.

Euonymus Linn. Celastraceae

Burning bush, Golden bush, Euonymus

Plants are shade adapted and attain heights of up to 3m. A number of species that are deciduous to evergreen are grown for their ornamental fruits and foliage, which attain attractive purple-red autumn tints. *Euonymus japonicus* Wall. has coloured foliage and yellow flowers. *Euonymus sachalinensis* Maxim. has red to purple fragrant flowers. Plants are popularly grown as hedges.

Forsythia Vahl. Oleaceae

Golden Bells, Forsythia

Plants are hardy, shade tolerant, deciduous and attain heights of up to 3m. They are spring flowering. Flowers are pale yellow in *Forsythia suspense* Vahl. Called the Weeping Golden Bells and are bright yellow in *Forsythia viridissima* Lindl. called the Green stem Forsythia.

Fuchsia arborescens Onagraceae

Hanging Dewdrop

Plants are shrubs with spreading habit. The flowers droop gracefully and are fragrant and coloured pink to mauve-purple. *Fuchsia fulgens* is shorter in stature and has yellow to orange blooms.

Magnolia pumila Magnoliaceae

Magnolia

Plants attain heights of up to 4m. They are shade loving. The flowers are fragrant, drooping and creamy white in colour and are produced in spring.

Nandina domestica Linn. Berberidaceae

Heavenly bamboo, Nandina, Sacred bamboo

Plants are shade tolerant clump forming evergreens attaining heights of up to 3m. The foliage is ornamental and purple to red tinged. Plants are summer flowering. The white flowers are borne in large clusters. Propagated by seeds and division of old clumps.

Paeonia suffruiticosa Andr. Ranunculaceae

Peony

Plants are deciduous and attain heights of up to 1.5m. The flowers are single to double and coloured white, pink, red or yellow. They resemble roses.

Rhododendron Linn. Ericaceae

Azalea, Rhododendron

Plants have a spreading habit and species include those that are tree-like such as *Rhododendron albrechtii, Rhododendron arborescens* and *Rhododendron atlanticum* and those that are short in stature like *Rhododendron indicum* Sweet. and *Rhododendron kaempferi* Planch. Plants are mostly evergreen. The flowers are white to pink-mauve in colour. Evergreens are grown as hedges.

Spirea Linn. Spiraceae

Spirea

Plants have a spreading habit and attain a height of 3m.They are deciduous. Flowers are white to pink in colour with brown tipped anthers and are produced in summer. They can be grown in partial shade and are ideal for massing and for hedges. In *Spirea japonica* the leaves are variegated and the flowers are pink or red. *Spirea betulifolia* Pall. is a dwarf species with white flowers.

Syringa Linn. Oleaceae

Lilac

Plants are deciduous and attain heights of up to 3m. The flowers are pink or purple-mauve and are produced in spring and summer. *Syringa persica* is grown in the hilly regions of erstwhile Persia. *Syringa vulgaris* is the common Lilac, which bears fragrant purple-blue flowers in terminal spikes. Plants are propagated by seeds, hardwood cuttings and layers. A number species are grown in different parts of the world.

Tamarix gallica Linn. Tamaricaceae

Tamarix

Plants are deciduous arboreal spreading shrubs having ornamental foliage. The flowers are pink and are produced in summer. *Tamarix dioica* Roxb. has drooping branches and *Tamarix juniperina* Bunge. has feathery branches. Plants are propagated by seeds and cuttings.

Viburnum Linn. Caprifoliaceae

Viburnum

These are shrubs grown primarily for their ornamental foliage and fragrant flowers. Plants attain heights of up to 3m and have a spreading habit. They flower during the spring and summer seasons. *Viburnum sargentii* and *Viburnum trilobum* have purple-red foliage. *Viburnum opulus* and *Viburnum sieboldii* have creamy white flowers.

FOLIAGE SHRUBS

Monocotyledonous Foliage Shrubs

Nolina recurvata. Ruscaceae
syn. *Beaucarnea recurvata*

Pony tail tree/ Elephant foot tree

Plants are evergreen arboreal shrubs with swollen rounded stem bases that attain a width of more than 1.5m. The bark becomes longitudinally cracked with age. They are hardy, sun loving and suited for rockeries and for specimen planting. Leaves are lanceolate and up to 1m or more in length.

Cordyline Comm. ex Juss Laxmanniaceae/ Lomandraceae

Cordyline

Plants are evergreen shrubs attaining the stature of small trees. They are adapted to open and partially shaded situations. The stems branch rarely. They are rhizomatous and have unusually coloured leaves. The lanceolate leaves are sessile and have unforked leaf bases that distinguish them from Dracenas. They have been grouped by some systematists in the segregate families of *Lomandraceae* or *Laxmanniaceae* which include other selected genera. The cultivated varieties are grouped into the following species: -

Cordyline australis Hook. Laxmanniaceae / Lomandraceae

syn. *Cordyline indivisa / Cordyline terminalis*

Red Ti palm

Plants are hardy shrubs having spectacularly coloured bright pink leaves that turn maroon-red with age. They are very popular garden plants and are ideally suited for brightening indoor areas.

Cordyline fruiticosa Goepp. Laxmanniaceae / Lomandraceae

Ti palm, Palm lily

Plants provide attractive colour when group planted in gardens or when displayed individually in pots. The leaves are coloured red and green.

Cordyline rubra Hueg. ex Kunth Laxmanniaceae / Lomandraceae

Cordyline

The leaves are maroon-red and cupped in these. Plants are ideal for mixed planting with other species.

Dracena Vand. ex Linn. Dracaenaceae

Dracena

Plants are hardy trees, shrubs or herbs that thrive in open and partially shaded situations. The leaves are narrowly lanceolate to broadly lanceolate and are arranged spirally and appear clustered around the stem. The clasping leaf bases are forked like a fish tail or are wedge shaped. Plants previously grouped under *Pleomele* are now included in this genus. Cultivated varieties are grouped under the following species: -

Dracena americana Dracaenaceae

American Dracena

The lamina is shiny green, up to 30cm in length and appears flaccid. The stem is marked with cane-like markings of fallen leaf scars. Shoots are found to branch from the base upwards.

Dracena demerensis Dracaenaceae

Corn plant

Leaves are reflexed and have smoky- white and green bands on the lamina. The subtle colourings of grey, green and white of the leaves make plants particularly attractive.

Dracena draco Linn. Dracaenaceae

Dragon tree, Dragon's Blood tree

Plants are indigenous to the Canary Islands. They are massive, with thickened and bulging stems attaining a height of up to 15m. Leaves are more than 0.5m in length, dark maroon-red in colour and are held in terminal clusters on spectacularly branched stems. The leaf bases and plant sap are orange-red in colour and the latter is used to colour varnishes.

Dracena fragrans Ker-Gawl. Dracaenaceae

Fragrant Dracena, Palmillo

Leaves are broadly strap-like and reflexed, with a yellow band on the lamina. The flowers are pinkish-white, fragrant and are produced in axillary clusters. A very attractive plant suited for specimen planting in small partially shaded gardens. Plants are ideal for pot display too.

Dracena goldieana Dracaenaceae

Saucer-leaf Dracena

Plants have ovate leaves that are held horizontally. The lamina has acuminate tips and have regularly spaced golden yellow patches on a green base.

Dracena marginata Dracaenaceae

Narrow leaved Dracena

Leaves are narrowly lanceolate, long and arching with pink lines. The stems are rarely branched. They are displayed as specimen plants in small gardens and included in rockeries.

Dracena reflexa Dracaenaceae
Syn. *Pleomele reflexa*

Song of India

The leaves are up to 15cm long, slightly twisted and are spirally arranged with or without creamy yellow longitudinal bands on the lamina.

Dracena sanderiana hort. Sander ex Mast. Dracaenaceae

Lucky Bamboo

Plants are shade adapted and tall growing shrubs having variegated leaves with sheathing leaf bases that clasp the stems. Older stems from which leaves have shed or are stripped off, resemble bamboo canes. Such one to two noded stem cuttings that have regenerated shoots are popularly grown indoors in water and are called Chinese Lucky Bamboos.

Dicotyledonous Foliage Shrubs

Acalypha wilkesiana Muell. Arg. Euphorbiaceae
Syn. *Acalypha amentacea* Roxb./ *Acalypha tricolor* Seem.

Vareigated Acalypha

Plants are evergreens attaining a height of more than 4m. The shoots and leaves are succulent. The leaves are ovate to cordate with the lamina often being pleated. The leaves are green or coppery red with creamy yellow or maroon to pink patches and have serrate margins. Plants are ideal for partially shaded situations especially for hedge making.

Aphelandra squarrosa Nees Acanthaceae

Zebra plant

Plants attain a height of up to 3m. The leaves are decussate, elliptic to lanceolate and have clearly marked yellow veins. They are suited for growing in shaded and partially shaded situations in the tropics and can be grown in the open at higher elevations. Older plants flower readily under both situations. The flowers are yellow and are protected by maroon-red bracts.

Codiaeum vareigatum Euphorbiaceae

Crotons

Plants are sun-loving evergreens that are popularly grown in tropical gardens. Cultivated varieties include small leaved sub-shrubs to arboreal shrubs. The leaves are brilliantly coloured and variously shaped. The lamina is entire to lobed and curly, cupped or drooping. Plants are sturdy and need yearly pruning to control overgrowth.

Eranthemum nigrum Linden Acanthaceae

Eranthemum

Plants are evergreens attaining a height of up to 1.5m. The shoots and leaves are entirely purple in colour. Plants are adapted to partial shade. The leaves are elliptic, glossy or dull with smooth margins and acute tips.

Euphorbia cotinifolia Linn. Euphorbiaceae

Butterfly plant, Red Spurge

Plants are thin-stemmed sub-shrubs attaining heights of up to 1m. The leaves are ovate to lobed like the wings of butterflies and are maroon red to reddish green in colour. Plants are sparsely leaved and semi perennial in habit.

Euphorbia drupifera Thonn. Euphorbiaceae
Syn. *Elaophorbia drupifera*

Euphorbia

Plants are arboreal shrubs attaining a height of up to 15m and are indigenous to tropical and southern Africa. They have been introduced into Indian gardens as succulent foliage plants. The young shoots are rounded and carry clusters of thick obovate leaves that are attractive. The shoots are armed with pairs of spines.

Euphorbia pulcherrima Willd. ex Klotzsch Euphorbiaceae
Syn. *Poinsettia pulcherrima*

Poinsettia, Christmas flower, Mexican Flame leaf

Plants attain a height of up to 2m. Inflorescences (cyathia) are produced terminally during the cooler months of December-January. These are subtended by clusters of red and green coloured leafy bracts.

Gmelina philippinesis Cham. Lamiaceae
Syn. *Gmelina hystrix* Schult.

Hedgehog plant, Wild sage

Plants attain heights of up to 3m and are partially shade loving. The shoots are pendant to climbing in habit. Inflorescences are pendant with a length of up to 30 cm. The flowers are yellow and cup shaped with four petal lobes. They are enclosed in tan-coloured bracts.

Graptophyllum pictum Griff. Acanthaceae

Caricature plant, Graptofilo

Plants belong to a monotypic genus indigenous to N.E. Australia and N. Guinea. They are evergreen attaining a height of up to 2m. The broadly ovate leaves have splashes of creamy white or pinkish colour centrally on the lamina. The leaf blades are slightly puckered and the margins are undulate. Cultivar 'Goldiana' has green with splashes of golden yellow on the leaves.

Hoffmannia refulgens Hemsl. Gesneriaceae
Syn. *Hoffmannia ghiesbreghtii* Hemsl.

Hoffmania, Strawberry Splash, Tafetta plant

Plants are shade loving and have attractive foliage. The leaves are thick, lanceolate, 15 to 25 cm long; coloured reddish-green and have a felt-like surface texture. The lamina appears quilted due to the depressed veins.

Lawsonia inermis Linn. Lythraceae
Syn. *Lawsonia alba* Lam.

Henna, Mignonette plant, Mylanchi

Plants attain a height of up to 3m in gardens. They are adapted to partially lighted as well as open situations. Flowers are creamy white and delicately scented. The leaves are small, obovate and are held on thin and arching branches. Plants were previously popular hedge plants of gardens.

Leea guineense G. Don Leeaceae
Syn. *Leea coccinea* Planch.

Leea

Plants are adapted to partial shade and attain a height of up to 4.5 m. Leaves are glossy, bipinnately compound with sparsely toothed margins towards the upper half. They are green with maroonish-black coloured shoots. Inflorescences are cymes produced terminally and carry small pink flowers with a red calyx.

Manihot esculenta Crantz. cv. 'Vareigata' Euphorbiaceae
Syn. *Manihot utilissima* Pohl.

Vareigated Manioc, Vareigated Tapioca

This variegated-leaved selection of Tapioca is grown in gardens as a specimen plant. The leaflets are edged dark green and have a bold splash of creamy yellow colour along the center.

Muehlenbeckia platyclados Meissn Polygonaceae
Syn. *Homalocladium platycladium*

Ribbon bush, Tape worm plant

Plants attain heights of up to 3m and are adapted to shady situations. The leaves are small, lanceolate and are shed before flowering. The shoots are flattened phylloclads with distinct nodes. The flowers are minute and arise in clusters along the nodes.

Phyllanthus rosea Euphorbiaceae

Phyllanthus

These are small leaved shrubs attaining heights of up to 2m in gardens when grown singly. They are popularly grown as foliage hedges and made into topiary of different forms in the tropics. *Phyllanthus nivosis* Bull. has larger rounded leaves.

Pittosporum tobira Dryand Pittosporaceae

Mock Orange, Japanese Pittosporum

Plants are arboreal shrubs equally well adapted to partial and full sunlight. The leaves are obovate, 5 to 7 cm long with leathery and glossy lamina. They are held in terminal whorls on the branches. The variegations on the leaves are grayish green and white in colour.

Polyscias Forst. Araliaceae

Plants are excellent shade loving foliage ornamentals that have restricted branching and can withstand pruning. Members were previously classed in the genus *Aralia.*

Polyscias balfouriana Araliaceae
Syn. *Aralia balfouriana / Panax balfourii*

Dinner Plate Aralia

Plants are bushy shrubs adapted to partial shade, having a dense cover of rounded trifoliate leaves that attain a diameter of up to 10 to 15cm. The leaflets are coarsely toothed, shiny and bright green in colour. In cultivar 'Marginata' the leaflets have irregular white variegations along the margins. In cultivar 'Pennockii', the leaflets have patches of pale green and creamy white colour on the lamina.

Polyscias filicifolia Araliaceae
Syn. *Polyscias filicifolium / Panax filicifolius*

Fern-leaved Aralia

Plants have succulent young shoots and petioles that carry finely cut fern-like bipinnate leaves. The mid rib is tinged purple. Plants are ideal for pot growing in shaded situations.

Polyscias fruticosa Araliaceae
Syn. *Aralia fruiticosa / Panax fruiticosum / Nothopanax fruiticosum*

Ming Aralia, Parsley Aralia

Plants are erect in habit with small leathery bi-pinnate leaves that have margins cut into segments. The young shoots and petioles are spotted white. In a variegated cultivar, the leaflets have a white edging.

Polyscias guilfoylei Araliaceae
Syn. *Aralia guilfoylei / Nothopanax guilfoylei*

Black stemmed Aralia, Frosted Aralia, Geranium Aralia

Plants are shade loving and have pinnate leaves of up to 40 cm length. The pinnae are lobed with irregularly toothed margins.

Polyscias paniculata Baker Araliaceae

Variegated Aralia

Plants are erect in habit. The pinnae are ovate with serrate margins. The lamina has variegations of creamy yellow and various shades of green. Plants were previously very popular foliage plants of homesteads.

Polyscias scutellaria Araliaceae

Round leaved Panax, Shield Aralia

Plants are adapted to shady situations. They are bushy with closely spaced shoots and leaves. The leaves are rounded, up to 4cm in diameter and are bright green with broadly toothed margins. They are ideal for topiary in pots.

Pseuderanthemum atropurpureum Acanthaceae

False Eranthemum

Plants are shade-adapted shrubs attaining heights of up to 2m or more. Cultivated varieties have leaf variegations of light to dark maroonish purple. The flowers are light pink in colour. *Pseuderanthemum carruthersii* has grayish white leaves and the flowers are white.

Pseuderanthemum reticulatum Acanthaceae
Syn. *Pseuderanthemum carruthersii* var. *reticulatum*

Golden Net bush

Plants attain a height of up to 2m. The leaves are broadly ovate, yellowish green with darker green veins. The leaves darken with age. The flowers are produced terminally on the shoots. They are white with magenta-pink spots.

Schefflera Forst. Araliaceae

Schefflera

This is a genus of evergreen arboreal shrubs or trees adapted to partially lighted and open situations.

Schefflera arboricola Araliaceae
Syn. *Heptapleurum arboricolum*

Dwarf umbrella tree, Dwarf Schefflera

Plants are arboreal evergreens attaining heights of more than 3.5m. The leaves are palmately compound and carry 7 to 10 leaflets of 10 to 15 cm length. A variegated cultivar having creamy yellow patches on the leaves is also grown. They are very popular indoor plants.

Schefflera elgantissima Araliaceae
Syn. *Aralia elegantissima /Dizygotheca elegantissima*

Elegant Schefflera, False Aralia

Plants are arboreal shrubs attaining heights of more than 4m under shaded situations. The shoots are erect and they carry palmate leaves that have narrow leaflets with broadly toothed margins. The leaves are coloured dark green.

Senecio cineraria DC. Solanaceae
Syn. *Cineraria maritima / Senecio candicans*

Dusty Miller

Plants are semi-perennial sub shrubs attaining a maximum height of up to 1 m. They are grown in the open in sub-tropical and temperate situations and are being grown under shade in the humid tropics. The leaves are small, ovate, lobed silver-grey in colour. The flowers are small, yellow and are produced in flat-topped terminal cymas. For ideal foliage renewal, plants are to be replanted every two years.

contd. colour plates

Allamanda cathartica

Allamanda cathartica 'Double'

Allamanda 'Cherries Jubilee'

Allamanda schottii

Kopsia fruticosa

Neerium oleander

Tabernaemontana coronaria

Tabernaemontana coronaria 'Flor-Pleno'

Tabernaemontana indica

Tabernaemontana indica 'Nana'

Tabernaemontana indica 'Silver Rain'

Vinca rosea

Bauhinia acuminata

Caesalpinia pulcherrima

Calliandra haematocephala

Carissa carandas

Cassia biflora

Quassia amara

Asystasia travancorica

Barleria cristata 'Rosea'

Odontonema cuspidatum

Pachystachys lutea

Pseuderanthemum reticulatum

Ruellia formosa

Beloperone guttata

Beloperone lutea

Dianthera nodosa

Jacobinea carnea

Schaueria flavicoma

Thunbergia erecta

Acalypha hispida

Argyreia speciosa

Jacobinea coccinea

Jatropha hastata

Jatropha podagrica

Phlebophyllum kunthianum

Clerodendrum fragrans

Clerodendrum macrosiphon

Clerodendrum nutans

Clerodendrum speciosum

Clerodendrum squamatum

Clerodendrum thomsoniae

Hibiscus mutabilis

Hibiscus rosasinensis 'Double'

Hibiscus rosasinensis 'Flor-Pleno'

Hibiscus rosasinensis 'Hose-in-hose'

Hibiscus syriacus

Malvaviscus arboreus

Ixora acuminata

Ixora chinensis

Ixora coccinea

Mussaenda erythrophylla 'Rosea'

Mussaenda glabra

Pentas lanceolata

Gardenia jasminoides

Hamelia patens

Podranea ricasoliana

Randia maculata

Tecoma stans

Tecomaria capensis 'Flava'

Bougainvillea 'Mahara Pink'

Bougainvillea 'Thimma'

Galphimia gracilis

Ochna kirkii

Otacanthus caeruleus

Tibouchina elegans

Aglaonema costatum

Dieffenbachia maculata

Oscimum sanctum

Philodendron 'Emerald Duke'

Platycladus orientalis

Syngonium podophyllum

Dracena deremensis

Dracena fragrans

Dracena godseffiana

Dracena goldieana

Dracena marginata

Dracena sanderiana

Aphelandra squarrosa

Duranta variegata

Graptophyllum pictum

Graptophyllum pictum 'Goldiana'

Pseuderanthemum atropurpureum

Pseuderanthemum carruthersii

Cordyline rubra

Cuphea hyssopifolia

Cuphea hyssopifolia 'Alba'

Polyscias balfouriana 'Variegata'

Polyscias fruticosa

Polyscias scutellaria

Acalypha godseffiana

Codiaeum variegatum

Hibiscus tiliaceus 'Variegata'

Manihot esculenta 'Variegata'

Murraya exotica

Phyllanthus rosea

Chapter 3

FLOWERING AND FOLIAGE HERBS

Herbs include non-woody or succulent-stemmed plants that are relatively shorter in stature when compared to shrubs. Based on their ornamental nature they can be broadly grouped as **Flowering Herbs** and **Foliage Herbs. Flowering Herbs** can be further classified based on their duration into **Annual Flowering Herbs** and **Perennial Flowering Herbs**. Foliage herbs of a perennial nature are further differentiated into monocotyledonous and dicotyledonous foliage herbs.

ANNUAL FLOWERING HERBS

Annual herbs are grown in gardens for their spectacular beauty when they are in flower. They are seasonal in habit and have duration of up to four months. Most of them have a short vegetative phase, which is followed by continuous flowering for several weeks. They are used in gardens for obtaining various colour effects. Varieties of annuals having analogous or similar flower colours can be grouped in flowerbeds and flower borders for monochromatic effects. For contrasting colour effects, varieties having complementary or contrasting colours like violet and yellow or blue and orange as well as green and red can be used. For colour blending effects, flowers having different shades of the same colour can be grown.

CLASSIFICATION OF ANNUAL FLOWERING HERBS

Annual Flowering herbs are broadly classified into cool season annuals and warm season annuals.

COOL SEASON ANNUAL FLOWERING HERBS

These are adapted to grow and flower in temperate latitudes. They have a remarkable range in size, colour and bloom characteristics. The requirement of low temperatures for germination and the tolerance or intolerance of frost by seedlings determines when and where these annuals are to be sown for adequate and timely flowering. They can further be subdivided into the following groups:

a) Late winter-sown early-blooming annuals

These are hardy annuals, sown late in winter inside green houses and the seedlings are hardened and transplanted outdoors in spring. Alternately their seeds can be sown outdoors early in spring as soon as the soil is workable. These are generally early bloomers. Examples include *Clarkia*, Cornflower, *Cynoglossum*, *Godetia*, *Gypsophila*, Larkspur, *Linaria*, *Myosotis*, Poppy, *Rudbeckia*, Snap Dragon and Sweet William. Heliotrope is an annual that needs to be sown indoors in mid winter itself for transplanting in spring.

b) Late-blooming annuals

These are annuals that are slow growing during their seedling stage and are hence to be germinated earlier under protected environments to enable them to complete a considerable portion of their vegetative phase before planting out. This enables them to commence flowering early and to complete flowering before the onset of winter. Examples include *Browallia*, California Poppy, *Celosia*, *Dyssodia*, Morning Glory, *Salpiglossis* and *Tithonia*.

c) Frost-intolerant annuals

The seedlings of these annuals are susceptible to frost and are hence sown indoors and seedlings are transplanted when the danger of frost has passed. Examples include *Acroclinum*, *Arctotis*, *Ageratum*, *Ammobium*, *Anchusa*, Aster, Carnation, Cape daisy, *Cleome*, Flax, *Gaillardia*, Hollyhock, Ice plant, *Molucella*, *Phlox*, Pinks, *Scabiosa*, *Schizanthus*, Statice, Stock, Strawflower and *Trachymene*.

d) Direct-seeded or *in situ* sown annuals

These are annuals that cannot withstand transplanting and are hence sown *in situ* and later thinned as per requirement. Examples include Larkspur, *Linaria*, Poppy, Nasturtium, *Nemisia*, *Nemophila*, *Nicotiana* and *Nigella*.

WARM SEASON ANNUAL FLOWERING HERBS

These are tolerant to high temperatures of the tropics and are further grouped as follows:

a) Summer flowering annuals: Examples include *Cosmos*, *Coreopsis*, *Gaillardia*, Sunflower, *Tithonia*, *Torenia* and *Zinnia*.

b) Rainy season annuals: These are annuals that can withstand heavy showers after flowering. Examples include Balsam, *Cosmos*, *Gomphrena*, Marigold, Sunflower and *Tithonia*.

c) Non season bound annuals for the humid tropics

   A few annuals can flower in the humid tropics as they do in temperate latitudes. However, in the humid tropics, certain high temperature-tolerant annuals can be grown throughout the year. Examples include Balsam, *Cosmos*, *Gomphrena*, Marigold, *Tithonia* and foliage Amaranths.

d) Cool-season annuals for the humid tropics

   Certain cool season annuals that have been started elsewhere can be grown in the relatively cool season from early November to late January when night temperatures as low as 18 ^{0}C to 20 ^{0}C and day temperatures of 26 to 28 ^{0}C prevail. These include *Celosia*, *Coreopsis*, *Gaillardia*, *Gazania*, French marigold, *Phlox* and *Verbena*.

The popularly grown **warm season and cool season annual flowering herbs** of gardens of the tropics and the temperate regions are listed below:

Abronia umbellata Nyctaginaceae
Syn. *Abronia rosea*

Sand Verbena

Plants have a trailing habit. The flowers are mauve to lilac and scented. They resemble verbenas. Plants are popularly grown in rockeries and trailed over rocky slopes.

Acroclinum roseum Asteraceae
Syn. *Helipterum roseum*

Acroclinum, Immortelle, Everlasting flower

Plants are slender stemmed and attain heights of over 0.5m. The leaves are linear-ovate. The flowers are white, yellow or pink and have a papery texture. They resemble asters but are smaller. They are suited for use as dry flowers but are more delicate than straw flowers. They are summer flowering in temperate situations and winter flowering in sub tropical regions.

Adonia aestivalis Ranunculaceae

Pheasant's eye

Plants grow to a height of up to 40cm and prefer open situations. The leaves are finely cut and fern-like. The flowers are dark red, cupped and 2.5cm in diameter. They are spring flowering.

Ageratum conyzoides Asteraceae
A. houstonianum & *A. mexicanum*

Floss flower, Ageratum

Plants belong to a genus of garden annuals that are short in stature and hardy. They are dome shaped. A few species have been hybridized to evolve a number of hybrids having long lasting flowers. Flowers are fuzzy and their colours range from white to dark mauve. Plants are adapted to open sunny situations. They can be grown throughout the year.

Agrostemma githago Caryophyllaceae

Corn cockle

Plants are tall hardy annuals attaining heights of up to one meter. The flowers are pale lilac and often dotted with a darker shade. They are spring flowering.

Althaea rosea Malvaceae
syn. *Alcea rosea*

Hollyhock

Plants are tall, attaining heights of up to 2.5m. The large and rounded leaves are held away from the main stem and in their axils, all along the stem, single to fully double flowers in varying shades are arranged. They are summer flowering and seeds are to be sown immediately after winter.

Alyssum maritimum Brassicaceae

Sweet Alyssum, Mad wort, Sweet Alison

Plants have a mat like spreading habit with grayish green leaves. The flowers are delicate, white and honey scented. In mild sub tropical situations plants can be grown throughout the year.

Amaranthus caudatus Amaranthaceae

Love-lies-bleeding / Summer Poinsettia

Plants are tall growing attaining heights of up to 1.5m. The drooping tassel-like flower spikes are pale greenish yellow to crimson-red in colour. In less fertile soils vegetative growth is less and the flowers are better coloured.

Amaranthus hypochondriacus Amaranthaceae
Syn. *Amaranthus hybridus* var. *hypochondriacus*

Princes' feather

Plants attain heights of up to 1m. The foliage is green. The upright spikes are held like plumes above the level of the foliage.

Amaranthus tricolor Amaranthaceae
Syn. *Amaranthus melancholicus* var. *splendens*

St. Joseph's Coat

Plants have brightly coloured red, bronze and yellow leaves. They are suited for the humid tropics.

Amaranthus salicifolius Amaranthaceae

Chameleon, Fountain amaranth

Plants attain a height of up to 1m. The leaves are brilliantly coloured orange, bronze and red and are undulating and droop downwards gracefully.

Amaranthus bicolor Amaranthaceae

Summer Poinsettia

Plants have dark maroon foliage. The last few leaves produced terminally have a contrasting pink colour. The unusual colouring makes them ideal for pot display in gardens. Sowing can be done throughout the year in subtropical situations. In temperate regions planting should be done early in summer to get good colour in foliage.

Anchusa capensis Boraginaceae

Alkanet, Bugloss, Cape Forget-me-not, Summer forget-me-not

Plants are short in stature attaining heights below 0.5m with erect pubescent shoots. They are native to South Africa.The flowers are dark blue with a white colouration at the center, or are fully white in colour. *Anchusa italica* (syn. *Anchusa azurea*) is native to the Caucasian mountains of Europe and attains a height of up to 1.5m. The azure-blue flowers of this erect plant are attractive.

Antirrhinum majus Scorphulariaceae

Antirrhinum, Snapdragon

Tall to dwarf bedding types of this annual herb are grown in gardens. Flowers are single to double and their colours vary from red to mauve, maroon, orange, white and yellow, with or without variegations. The flowers with their prominent lip resemble the mouth of a dragon. Progenitor species of these garden annuals are perennials and are also grown in gardens

Arctotis grandis Asteraceae
Syn. *Arctotis stoechadifolia*

African daisy, Blue-eyed daisy

Plants attain heights of up to 0.5m and have a bushy habit. Flowers are coloured in pastel shades of buff, yellow, salmon, peach, off-white and lilac. The petals close during the night. The leaves are grayish green.

Brachycome iberidifolia Asteraceae

Swan river Daisy

Plants are dwarf thin stemmed annuals attaining heights below 0.5m. The flowers are fragrant white, blue, mauve-purple, violet or pink. Seeds are to be sown in situ.

Browallia demissa Solanaceae
syn. *Browallia elata*, *Browallia americana*

Browallia

Plants attain heights of up to 0.5m. Flowers are blue, violet or white and have a velvety texture. They are suited for hanging pots.

Cacalia coccinea Asteraceae
Syn. *Emilia flammea*

Tassel flower, Floral paintbrush

Plants are hardy with thin, wiry and hairy stems. The flowers are orange-red, yellow or white in colour and are borne in clusters on thin stalks like tassels. Plants thrive in open situations.

Calceolaria hybrida Scorphulariaceae

Calceolaria, Pocket-book plant, Pouch flower

Plants are grown as annuals and are herb-like. The flowers are highly coloured and pinched. Cool moist situations and shade are necessary.

Calendula officinalis Asteraceae

Calendula, Pot marigold

Plants are very popular as they can be grown almost throughout the year. The zinnia-like flowers are pale yellow to bright orange.

Callistephus chinensis Asteraceae

China aster, Aster

Plants are popular cut flower annuals that attaining heights of up to 1m. popularly grown cut flower varieties of India are white, mauve-purple or violet. Varieties grown in gardens are double to pom-pom types in flower form with a range of colours.

Campanula medium Campanulaceae

Canterbury bells

Plants are annuals or biennials having bell shaped flowers that are white, pink, mauve or violet-blue and rarely saucer shaped. They thrive in partially shaded situations.

Celosia argentea Amaranthaceae
syn. *Celosia plumosa*

Cock's comb

Plants grow to a height of 0.5 to 2m. The flower spikes may be feathery (plumose) or flattened (crested). Seeds are to be sown in spring for flowers in summer and are to be sown in summer for flowering during the rains. The inflorescences are brilliantly coloured red, yellow, orange or pink. Plants are true summer flowering annuals. They are suited for the humid tropics too.

Centaurea cyanus Asteraceae

Corn flower, Blue bottle, Bluet, Bachelor's button, Rogged Robin.

Plants are hardy, tall growing, attaining heights of up to 1m. Flowers are characteristically blue. A range of flower colours such as purple, pink, maroon and white has been evolved by mutations. Seeds are to be sown *in situ*.

Centaurea moschata Asteraceae

Sweet Sultan

These are tall erect annuals that carry fragrant tassel-like flowers on long stalks. The flowers are white, mauve to purple, yellow or red. They resemble powder puffs. In *Centaurea americana* the flowers are thistle shaped with rounded tops and a bulbous base.

Chieranthus cheri Brassicaceae

Wallflower

Plants are shrubby annuals or biennials that are dwarf to tall in habit. Flowers are produced on tall erect stalks. They are sweet scented and have a wide range of colours like yellow, orange, red and purple. Plants require cool climates for flowering.

Clarkia elegans Onagraceae

Clarkia, Rocky Mountain garland

Plants are hardy annuals adapted to partially shaded situations attaining heights of up to one meter. The fragrant flowers held on long stalks are single or double and have ruffled petals. They are coloured white, salmon pink, dark red or purple. The seeds are sown *in situ*. *Clarkia pulchella* is shorter in stature and has single or semi double flowers.

Cleome *spinosa* Capparidaceae
syn. *Cleome pungens*

Spider plant, Giant spider flower

Plants are tall attaining heights of over one meter and are entirely spiny. The leaves have a pungent aroma. The scented flowers are held in large clusters. They have prominent exerted stamens. The petals are pink, dark rose or white in colour. The flowering season extends for more than 3 months. Plants are adapted to warm temperate as well as tropical climates. The slender fruits, which hang down, resemble spider's legs. They are suited for the humid tropics too.

Clianthus dampieri Fabaceae
syn *Clianthus formosus*

Lobster's claw, Parrot's bill, Glory Pea

Plants attain heights of up to 0.75m. The flowers are showy, pendulous claw or beak like and coloured dark red with a black 'eye' at the base. Seeds have a hard seed coat and are usually sown *in situ*. They require chilling temperatures for flowering.

Collinsia bicolor Scorphulariaceae
syn. *Collinsia heterophylla*

Collin's flower

Plants are hardy, slender and dwarf in stature. The flowers are held on long spikes and are lilac, white or salmon pink. Sowing is done *in situ* in mild temperate climatic situations during September-October.

Consolida regalis Ranunculaceae

Annual Larkspur

Plants are branching or non branching and have bright green lacy leaves. The spikes carry purple, blue, rose or white flowers.

Convolvulus L. Convolvulaceae

Morning glory/ Dwarf Morning glory

Plants are bushy and have blue, pink, lilac or red flowers. They are day bloomers.

Coreopsis tinctoria Asteraceae
Syn. *Coreopsis elegans*

Tick-seed

This species is widely grown in gardens of temperate regions. The flowers are yellow and dark maroon-red. Plants may attain heights of up to 1m. They are suited for the humid tropics too.

Coreopsis drummondii Asteraceae
Syn. *Coreopsis diversifolia*

Coreopsis, Calliopsis

This species has yellow flowers with dark red centers. Dwarf varieties and the tall ones are ideal for massing.

Cosmos bipinnatus Asteraceae

Cosmea, Cosmos, Mexican Aster

Plants are tall, graceful annuals attaining heights of over 2m. they are ideal for mass planting against fences and walls to obtain a colourful cover. The foliage is finely cut. Flowers are yellow, white, pink, light or dark purple or maroon in colour with widely serrate petals. They are suited for the humid tropics.

Cynoglossum L. Boraginaceae

Chinese Forget-me-not, Hound's tongue.

Plants attain heights of up to 0.5m. The flowers are small and are held in slender gracefully branching stems. They are blue, white or pink.

Dahlia variabilis Asteraceae

Dhalia

Plants are tuber-forming perennials that are grown as annuals in the tropics and in temperate regions. Shoots are succulent, erect and carry lobed leaves. They produce flowers in few-flowered terminal clusters. The numbers of growing shoots that flower at a time are limited by pruning to one to three. The flowered shoots are cut back when plants are grown as perennials. When grown as annuals, the flowered shoots are allowed to die back and the tubers are lifted and stored for the next season. In the humid tropics where continuous flowering can be obtained, vegetative shoot tips are rooted and used for planting season after season.

Delphinium ajacis Ranunculaceae

Larkspur, Hyacinth-flowered Larkspur

Plants attain heights of up to one meter. Flowers are held on erect pointed terminal inflorescence stalks. Flowers are white, blue, mauve, pink, red or violet in colour. Seeds are generally sown *in situ*. Other species grown in gardens include the Macedonian or Siberian Larkspur- *Delphinium paniculatum* is a native of the Balkans and the Chinese Delphinium- *Delphinium grandiflorum* syn. *Delphinium sinense* is native to China. Both these are dwarf in stature, attaining heights of up to 30cms.

Dendranthema (see under Warm season Perennial flowering herbs)

Dianthus caryophyllus Caryophyllaceae

Carnation

Plants are popular annual cut flowers that are weak stemmed and require adequate staking for proper flowering. The leaves are linear, lanceolate and grayish green in colour. Varieties are classified into several groups based on floral characteristics. Flowers are white, pink, yellow or red or different shades of these colours.

Dimorphotheca sinuata Asteraceae
Syn. *Dimorphotheca aurantiaca*

African daisy, Cape daisy, Cape marigold, Star of the Veldt

Plants are short in stature with slender stems attaining heights below 0.5m. They produce flowers that resemble daisies and are coloured white, yellow, rose pink, salmon pink, orange or red with a blue coloured underside. Flowers remain open on sunny days only.

Dyssodia Willd. Asteraceae

Dalhberg Daisy, Golden Fleece

Plants have aromatic finely cut leaves and bright yellow flowers.

Eschscholzia californica Papaveraceae

California Poppy

Plants are hardy perennials that are native to California. They are dwarf, hardy plants with finely cut foliage. The flowers are conspicuous with their bright yellowish orange saucer shaped flowers that have a metallic luster. They remain open during sunny hours only. California Poppy is the state flower of California.

Felicia bergeriana Asteraceae
syn. *Cineraria bergeriana / Aster bergerianus*

Kingfisher daisy

Plants are dwarf in stature attaining heights below 30cm. A large number of small blue coloured flowers having a metallic luster are produced. The flowers remain open during sunny hours.

Gazania splendens Asteraceae

Treasure flower, Gazania

Plants are low growing and have a semi trailing habit. The leaves are pubescent with a silvery surface colour. Daisy- like flowers with ray florets coloured white, yellow, orange, red or pink and having contrasting centers are produced. Flowers remain open during sunny periods only.

Gilia capitata Polemoniaceae

Gilia, Ipomopsis

Plants attain heights of up to 0.5m. The foliage is finely cut and ornamental. The flowers are small and mauve-blue in colour. Other species grown are *Gilia tricolor* syn *Gilia nivalis* called bird's eye having white, pink or lavender-blue flowers and *Gilia rubra* syn. *Gilia coronopifolia* having red flowers. All the above prefer sunny situations and the seeds are sown *in situ* for propagation.

Godetia grandiflora Onagraceae

Godetia amoena

Satin flower, Farewell-to-spring

Garden Godetias are hybrids of the above two species and have dwarf cultivars attaining heights below 30cm and taller ones attaining heights of up to 0.75m. Flowers are produced in spikes and are often scented, single or double and attractively coloured white, rose, red, orange or blue with or without blotches. The petals are satiny in texture. In very fertile soils plants tend to be leafy with lesser flowers.

Gomphrena globosa Amaranthaceae

Bachelor's button, Globe amaranth, Vadamalli (Tamil, Mal.)

Plants are short in stature, hardy and bushy producing small button like to globose flower heads coloured white, rose, magenta-pink or violet-pink. They are suited for the humid tropics. They are suited for use as dry flowers too.

Gypsophila elegans Caryophyllaceae

Baby's breath, Chalk plant

Plants are delicately branched annuals bearing a large number of very small flowers coloured white, pink or dark rose terminally. They are suited for massing. The flowering branches have a misty or dew-strewn appearance from a distance. Plants are harvested entirely and the flowering branches are used decoratively in the fresh and dried state. Seeds are sown *in situ* at intervals in temperate regions for obtaining flowers continuously.

Helianthus annuus Asteraceae

Sunflower

Plants are the tallest among annuals grown in gardens. Plants attain heights of up to 3m and have large coarse-textured leaves. The flowers are commonly golden yellow. Those that are shaded bronze-red, chocolate-red and maroon-red are also grown. *Helianthus debilis* syn. *Helianthus cucumerifolius* is the Japanese Sunflower that is short and bushy in stature producing numerous small yellow flowers in large numbers. They are suited for the humid tropics too.

Helichrysum bracteatum Asteraceae

Everlasting flower, Straw flower

Plants are dwarf thin-stemmed annuals. The flowers are coloured silvery white, golden yellow, bronze-red or salmon pink. They have a papery texture and a metallic sheen that are not lost on drying. The half-open flowers are cut for drying.

Helitropium arborescens Boraginaceae

Cherry pie, Heliotrope

Plants are perennials that are grown as annuals. They are short in stature attaining heights of up to 30cm. The flowers are small and are held in clusters above the level of the foliage. The flowers are white, blue or purple in colour and are vanilla-scented.

Iberis umbellata Brassicaceae

Candytuft

Plants are dwarf in stature. The flowers are borne in spreading umbels. They are coloured white, pink, rose, lilac or purple. Plants are typical temperate annuals that do not thrive in humid tropics.

Impatiens balsamina Balsaminaceae

Balsam, Touch-me-not, Gul mendi (Hindi), Dopati (Beng.) Kaasi Thumba (Mal.)

Plants are succulent and dwarf in stature. They bear single or double flowers close to the main shoot or in the terminal branches. The branches are few and upright in habit. Flowers are white, pink, purple or red and variously shaded and occasionally spotted or blotched. They are hardy and thrive even during the heavy rains of the tropics. The seed carrying pods tend to burst open at the slightest pressure when ripe. They are suited for the humid tropics.

Lathyrus odoratus Fabaaceae

Sweet Pea

Plants are bushy or trailing temperate region annuals that produce bonnet shaped sweet scented flowers in a range of colours from white to red. The petals are often frilled . They are popular cut flowers too. Being deep-rooted hard soils are to be well loosened by 'trenching' during the vegetative growth period for proper flowering later.

Limonium sinuatum Plumbaginaceae
syn. *Statice sinuatum*

Sea lavender, Statice, Thrift

Plants are hardy biennials that are grown as annuals and attain heights of up to 0.5m. The shoots are angular and carry a rosette of leaves that lie close to the ground. The flowers are white, rose, mauve, yellow or dark blue and are papery in texture. They are ideal for drying and decorative use. Other species grown are the yellow flowered *Limonium bonduellii* and the long stalked and pink flowered *Limonium. suworowii.*

Linaria maroccana Scorphulariaceae

Toadflax, *Linaria*

Plants are popular garden annuals of Europe. They are short and bushy. Flowers are small, white, yellow, pink, purple, blue or bicoloured with or without blotches on the lip.

Linaria alpina is a species that has violet-blue flowers with an orange colouration on the lip.

Linum grandiflorum Linaceae

Scarlet Flax

Plants attain heights of up to 0.5m. the foliage is grass like. The flowers are white, pink, blue or red with a dark colouring at the center. They are held loosely on terminal panicles. The common flax *Linum usitatissimum* has light blue flowers and is also grown in gardens.

Lobelia erinus Campanulaceae

Lobelia

Plants are dwarf in stature and weak stemmed. The flowers are small and are coloured white, pink, red or blue with a white or yellow eye. They are semi-perennials and are grown as annual bedding plants in temperate regions.

Lupinus lupinellus Fabaceae

Lupin

Plants are popular annuals and biennials of gardens. A large number of flowers are held compactly on terminal spikes. *Lupinus hartwegii* has flowers that are coloured white, pink or pale blue; *Lupinus lutens* has sweet scented yellow flowers, *Lupinus subcarnosus* the Texas blue bonnet is blue flowered and dwarf in stature and *Lupinus mutabilis* attains a height of up to 1.5m and has dark blue flowers with yellow or rose markings.

Matricaria eximia Asteraceae
Syn. *Chrysanthemum parthenium*

Matricaria, Feverfew, False Chamomile

Plants are low growing semi-perennials grown as annuals. The flowers are white or yellow, small and button like and double. They are ideal bedding annuals for mild temperate climates. The leaves are finely cut and aromatic.

Matthiola incana Brassicaceae

Stock, Gilly flower

Plants are favoured annual or biennial flowering plants of gardens of temperate regions. The flowers are small single or double and are held on long spikes. They are coloured white, pink, red or lavender or blue and are delicately scented.

Mesembryanthemum criniflorum Aizoaceae

Ice plant

Plants are low growing spreading annuals that have grass like leaves and bear daisy like flowers that are coloured white, lavender, pink, red, yellow or orange. They remain open during sunny hours only.

Mimulus tigrinus Scorphulariaceae

Monkey flower

Plants are short in stature attaining heights of up to 0.5m. They are semi-perennials grown as annuals. The flowers are two-lipped, up to 5.0cm in diameter and are coloured yellow, pink, or red and blotched with brown or maroon colours.

Mirabilis jalapa Nyctaginaceae

Marvel of Peru, Four-O' clock flower

Plants are bushy semi perennials that are usually grown as annuals in tropical gardens. They are shade loving and produce tuberous roots. The flowers are fragrant and are coloured yellow, white or pink-red. They open during the afternoon hours. They are suited for the humid tropics.

Molucella laevis Lamiaceae

Bells of Ireland, Molucca balm, Shell flower

Plants attain heights below 1m and produce tiny-cupped flowers that are white at the center. The green bell shaped calyx hangs on shoot tips. The calyx is persistent and has decorative value when dried too.

Myosotis alpestris Boraginaceae

Forget-me-not

Annual and biennial Forget-me-nots are widely grown in gardens. The flowers are very small, fragrant and coloured white, pink, light blue or dark blue with a yellow eye.

Nemesia strumosa Scorphulariaceae

Nemesia

Plants are generally dwarf in stature. They are used for edging and bedding in gardens of temperate regions. The flowers are cup like and white, blue, yellow, orange or red. They are borne in clusters terminally on long stalks. Seeds are sown *in situ*.

Nemophila menziesii insignis Hydrophyllaceae

Nemophila, Baby blue eyes, California Blue Bell

Plants are spreading annuals that produce small bell shaped flowers. The flowers are usually cup shaped, light blue coloured and have a white eye. Dark purple or white flowered plants are also grown. The foliage is pale green and pubescent.

Nicotiana alata grandiflora Solanaceae
Syn. *Nicotiana affinis*

Ornamental Tobacco, Flowering Tobacco

Plants are semi perennials that are grown in gardens as annuals. The sweet scented flowers are trumpet shaped and white, blotched with violet, yellow, red and green. They open during the evening hours.

Nierembergia caerulea Solanaceae
Syn *Nierembergia hippomanica*

Cup flower

Plants are semi perennials that are grown as annuals. The flowers are cup shaped and are violet-blue in colour with a yellow center. Plants spread well when grown as edges.

Nigella damascena Ranunculaceae

Love-in-a-mist, Devil-in-the-bush, Fennel flower

Plants are short in stature with finely cut foliage. The flowers are white, purple, blue or pink. Seeds are sown *in situ*. The flowers resemble cornflowers.

Oenothera biennis Onagraceae

Evening Primrose

Plants are tall biennials that can be grown as annuals. The flowers are yellow, large and open during the evening hours.

Papaver spp. Papaveraceae

Poppy

Plants belonging to this genus are characterized by cup shaped flowers. The species grown as garden annuals include the Tulip poppy *Papavar glaucum*, the perennial Iceland poppy *Papaver nudicaule*, the Oriental poppy *Papaver orientale*, the bicoloured and double flowered Corn poppy or Shirley poppy *Papaver rhoeas*, and the predominantly red flowered Opium poppy *Papaver somniferum*. Flowers are single or double. A wide range of flower colours is seen in presently grown garden varieties of these species.

Penstemon hybrida Scorphulariaceae

Beard Tongue

The annual Penstemon varieties grown in gardens of temperate regions are hybrids of several progenitor species. The flowers are tubular and are held on long spikes. They are coloured white, pink, purple or crimson.

Petunia hybrida Solanaceae

Petunia

Plants are colourful annuals grown in beds and window boxes. The flowers are trumpet shaped and have frilled petal margins. Single and double flowered varieties belonging to different horticultural classes based on the nature of flowering and flower sizes are popularly grown.

Phlox drummondii Polemoniaceae

Annual Phlox, Star flower

Plants are favoured annuals of temperate gardens. They have a compact habit. The flowers are delicately scented and are produced terminally in small clusters in profusion. They are coloured white, pink, red or purple with a contrasting or white eye.

Pimpinella monoica Apiaceae

Lady's lace

Plants attain heights of up to 1.5m and have finely cut foliage. The flowers are small white and are produced as large sprays.

Portulaca grandiflora Portulacaceae

Moss rose, Portulaca, Sun plant, Rock moss

Plants are low growing semi perennials that are grown as annuals. The flowers are single or double and are coloured white, yellow, pink or orange with or without streaks and spots. They are suited for the humid tropics.

Rudbeckia bicolor Asteraceae

Cone flower, Gloriosa daisy

Plants are hardy and tall growing, attaining heights of up to 0.5m. The flowers are daisy like and coloured yellow with a crimson-brown center. Plants favour semi-shaded and moist situations. *Rudbeckia hirta* called the Black-eyed Susan is biennial in habit and has large orange-red flowers with black centers.

Salpiglossis sinuata Solanaceae

Painted tongue, Velvet flower

Plants are winter flowering, tall annuals. They bear petunia like flowers profusely. The flowers are coloured white, pink, yellow, orange, red or violet. The petals have a velvety texture.

Salvia splendens Lamiaceae

Sage flower, Scarlet Sage

These are spectacular garden annuals that have flowers coloured scarlet, salmon pink or violet. Other semi-perennial species that are grown as garden annuals are *Salvia horminum* having blue, pink, or white flowers, *Salvia farinacea* having lavender-blue flowers *Salvia patens* with bright blue flowers and *Salvia coccinea* having small dark red flowers.

Saponaria calabrica Caryophyllaceae

Soapwort

Plants are short in stature. The flowers are small pink or white. A taller species attaining heights of up to one meter is*Saponaria vaccaria*.

Scabiosa atropurpurea Dipsaceae

Pincushion flower

Plants attain heights of up to 3m. Compact dwarf varieties are also grown. Flowers are large fragrant and are coloured lavender, blue, rose, red and coral red. They are covered by long silvery white stamens.

Schizanthus hybridus Solanaceae

Butterfly flower, Poor man's Orchid

Plants are fern like and have finely cut leaves. The flowers are delicate and two lipped resembling butterflies. The petals are bicoloured and spotted. Their predominantly pink, red or yellow flowers are veined and spotted too. Plants are cool season annuals.

Senecio cruentus Asteraceae
syn. *Cineraria cruenta*

Cineraria, Florist's Cineraria

Plants are shade loving annuals attaining heights of up to 0.5m. Flowers are daisy like and borne in clusters. They are coloured white, red, pink, purple or lavender-blue.

Tagetes erecta Asteraceae

African Marigold

Plants are hardy annuals having finely cut aromatic leaves. The flowers are light to dark orange or in varying shades of yellow. Cultivars ranging in heights from 0.5m to one meter are grown in gardens. They are suited for the humid tropics too.

Tagetes patula Asteraceae

French Marigold

Plants are bedding annuals of gardens. They are also popular potted plants. The flowers are smaller than those of African marigold. The ray florets maybe coloured orange or maroon-red and the flower heads are less compact than that of the other species. White flowered cultivars and those having non-aromatic leaves are highly prized. They are suited for the humid tropics too.

Tithonia rotundifolia Asteraceae

Mexican sunflower

Plants are hardy annuals attaining heights of 1.0 to 1.5m. The leaves are sunflower like. Flowers are reddish orange and up to 20cm in diameter. The flower stalks are brittle and hollow. They are suited for the humid tropics.

Torenia fournieri Scorphulariaceae

Wishbone flower

Plants are dwarf in stature attaining heights of 30 to 45cm. They are bushy and flower profusely unlike other annuals that thrive in the humid tropics. The flowers have a prominent lip. They are bicoloured, white tinted with pink, mauve and violet shades. Seeds can be sown *in situ* or transplanted.

Trachymene caerulea Umbelliferae

Blue Lace flower

Plants are slender stemmed with finely cut leaves. The flowers are borne in attractive umbrella shaped umbels coloured lavender blue, pink or white.

Tropaeolum majus Tropaeolaceae

Nasturtium, Indian cress

Plants have slender wiry stems. Tall plants have a climbing habit. The dwarfs are bushy in habit. The leaves are rounded. The flowers are spurred and are coloured yellow, pink, orange, red or maroon red with attractive spots and markings.

Ursinia anethoides Asteraceae

Jewel of the Veldt

Plants are native to S. Africa and adapted to cool winters. They are hardy and shrub like with finely cut leaves. The flowers are daisy like and are coloured dark orange with a dark red center. Two dwarf and low growing species that are also grown are *U. pulchra* and *U. pygmaea*. Both have bright orange flowers.

Venidium fastuosum Asteraceae

Namaqualand daisy

Another cool season annual of S African origin, plants are entirely pubescent and tall growing, attaining heights of up to 1m. The flowers are white to yellow coloured with dark red markings and a black center.

Viola hortensis Violaceae

Pansy

Plants are bienniels or semi perennials that are popular bedding annuals in gardens of the temperate regions. The butterfly like flowers with 5 overlapping petals have an uncommon blend of violet, yellow, dark rose and blue colours and is delicately scented.

Viola cornuta Violaceae

Viola, Tufted Pansy, Horned Violet, Bedding Pansy

Plants are tufted in habit and have smaller flowers than garden pansies. The flowers are coloured white, blue, purple, yellow, orange or red.

Zinnia elegans Asteraceae

Zinnia

Plants are popular annuals that are adapted to a wide range of climatic conditions. Flowers are single to fully double and available in a wide range of colours. Their range in height and the brilliance of their flower colours make them popular throughout the world. They are suited for the humid tropics too.

Zinnia haageana Asteraceae
syn.Zinnia *angustifolia*

The Persian carpet

Plants are low growing annual bedding plants having yellow and maroon red coloured flowers.

PERENNIAL FLOWERING HERBS

Perennial Flowering herbs are generally succulent stemmed and suited for sheltered or partially shaded situations in the humid tropics. They are ideal for massing in beds or pots. They require very light pruning, mainly for removal of aged leaves and trailing stems. They are classed into warm-season and cool season flowering herbs based on their climatic requirements for growth and flowering.

WARM-SEASON PERENNIAL FLOWERING HERBS

Popular **warm-season perennial flowering herbs** and **orchids** grown in Indian gardens are listed below.

Angelonia salicariifolia Plantaginaceae
syn. *Angelonia salicariaefolia* / *Angelonia angustifolia*

Angelonia / Narrow-leaf Angelon

Plants are herbs with succulent stems. The leaves are small, linear and ovate with glandular hairs. The shoots are aromatic. The flowers are slightly cupped with spreading lobes coloured white or purple. They are produced terminally on racemose inflorescences carrying leafy bracts. The flowered shoots are to be pruned regularly.

Anthurium schott Araceae

Tail flowers

Plants of this genus are shade loving creeping or climbing herbs with semi-epiphytic adaptations. The species commercially grown for flowers that have an erect to reclining spadix subtended by a colourful spathe include the Painter's Palette flower - *Anthurium andreanum* Lind. and *Anthurium amnicola* hybrids called the Miniature Anthuriums adapted to the humid tropics which are important cutflowers of tropical S. America, the Philippines, Mauritius and Sri Lanka and in India are confined to Kerala, the Coorg region and parts of West Bengal and the North- Eastern belt. *Anthurium scherzerianum* Schott. the Flamingo flower is adapted to the subtropics and is an important green house plant. The stems are thick, slow growing and carry closely spaced leaves in *Anthurium andreanum*. The heart-shaped lobed leaves of many of its cultivars are used for decorative purposes. Varieties bred for cut flowers have spathe colours ranging from pure white to pink, red, chocolate brown, orange, yellow and green with spectacular bicolors that combine green basal lobes with red, pink, peach or white shades in the rest of the spathe.

Aster amellus Asteraceae

Michaelmas daisy / Perennial Aster

Plants attain heights of up to 60cm. The leaves are ovate in shape. Plants sucker profusely after flowering. Flowers are single and daisy like and are carried on panicles. They are white, light blue or dark blue in colour.

Begonia semperflorens Begoniaceae

Flowering begonia

Plants are garden hybrids of unknown ancestry. The stems and petioles are often hairy. The leaves are obliquely ovate or sub-orbicular with a silvery shading on the upper surface. Flowers are unisexual, pink to white and are held erect or pendent in many flowered cymes. *Begonia nelumbifolia* called the 'lotus-leaved begonia' or 'lily pad begonia' has large rounded leaves with crenulated margins. These moisture-loving plants can be grown in partially-shaded open situations.

Centratherum intermedium Asteraceae

Pineapple leaf, Pineapple Sangria

Plants are succulent, dwarf spreading herbs having ovate leaves with serrate margins. The flowers are purple-mauve. They are ideal edging plants.

Chrysothemis pulchella Gesneriaceae

Chrysothemis, Copper-leaf, Square-stem

Plants are short statured succulent herbs producing root tubers, adapted to shade situations. The leaves are attractively coppery green with a thick and rough surface texture. The stems are light green. Bright yellow flowers enclosed in an orange coloured calyx and bracts are produced in leafy axils. Plants produce hairy tubers. Propagation is carried out with tubers or by herbaceous cuttings.

Coreopsis auriculata Asteraceae

Calliopsis, Coreopis, Tickseed

Plants are clump-forming perennials that attain a maximum height of 1.0m. They are evergreen and ideal bedding plants for sunny locations in gardens. The flowers are yellow, red or mauve-purple and are produced mainly in the summer months.

Coreopsis grandiflora is an erect perennial attaining a height of one meter to 1.2m. These are also summer flowering and the flowers are yellow or red in colour.

Dendranthema grandiflorum (Ramat.) Kitam. Asteraceae

Chrysanthemum, Florist's Mum

Plants are perennials but are usually grown as annuals. The shoots terminate in inflorescences. Cultivated varieties are grouped and grown as 'Standard mums' or 'Spray mums'. According to flower form, varieties are variously classified. After flowering, the suckers produced are separated for the next season's planting. In India, yellow flowered chrysanthemums are traditionally used for decorative purposes.

Dianthus chinensis Caryophyllaceae

Indian, Chinese or Japanese Pinks

Plants are semi- perennials that are thin stemmed or semi-decumbent in habit. They thrive in containers under shaded situations in the humid tropics. The leaves are small, grayish green, linear-lanceolate. The flowers are single, white or reddish pink with faint dark markings. The petals margins are finely fringed. They become straggly and show reduced growth with the onset of the hot season. Repotting every year after removal of old shoots is found to rejuvenate plants.

Impatiens wallisii Balsaminaceae
syn. *Impatiens hawkerii, Impatiens wallerana*

Perennial Balsam

Plants are succulent bushy semi-perennials attaining heights of 0.5 to 0.75m. The flowers are showy red, pink or white and with or without blotches of white or pink. They are ideal bedding plants for partially shaded areas.

Mirabilis jalapa Nyctaginaceae

Four-o'clock plant, Marvel of Peru

Plants are succulent semi-perennials thriving in semi-shaded situations, attaining heights below one meter. They have tuberous roots that help them to regenerate after the tops dry out in summer. The flowers are scented tubular and brightly coloured white, yellow, magenta, or pink with or without streaks of these colours. They open during late afternoon hours. The seeds are rounded and black. Plants propagated by seeds show variability in flower colour. Vegetative propagation is done by lifting the tubers after the tops dry out and re planting them before the rains.

Protulaca grandiflora Portulacaceae

Moss- rose, Rock rose, Sun plant, Purselane

Plants are semi-perennials grown as annual bedding plants in the open and in partially lighted situations in tropical gardens. They are entirely succulent and decumbent. The leaves are terete or ovate with rounded tips. The flowers are single to double. Flower colours range from white to light pink, magenta, orange and yellow often with streaks in the double varieties. Propagated by herbaceous cuttings.

Saintpaulia ionantha Gesneriaceae

African Violet, Usambara Violet

Plants are stem-less, clump-forming, finely hirsute and succulent perennials. They are shade loving and most ideal for growing in containers under protected situations. The leaves are dark green and have a velvety surface texture. The flowers are in varying shades of violet or white. Brilliantly coloured varieties are also presently grown in gardens. Propagated by whole leaf cuttings.

Salvia coccinea Lamiaceae

Blood Sage

Plants are succulent semi perennials that thrive in well-lighted partially shaded situations. The tubular flowers are carmine red. They attain a height of up to 0.5m. They are propagated by herbaceous cuttings.

Salvia superba Lamiaceae

Perennial Salvia

Plants are perennial in habit and are adapted to mild temperate situations. They are evergreen, bushy, erect and attain heights of up to one meter. The tubular flowers are red or purple.

Zinnia linnearis Asteraceae

Perennial Zinnia

Plants are low growing thin-stemmed bushy semi-perennials suited for open or naturally shaded situations, attaining heights of up to 45cms. The leaves are linear-ovate. The flowers are produced singly and have yellow or white ray florets. Plants are ideal for containers. To maintain them for more than one year they are to be protected during tropical summers.

Orchids Orchidaceae

In gardens of the humid tropics of India, a number of terrestrial, semi-terrestrial and epiphytic orchids are grown as ornamentals. From among the great diversity of genera and species found naturally and produced by multi generic and multi specific hybrid combinations, a handful that have shown adaptability and hardiness in varying climatic situations in the sub continent are also selectively grown. Due to their specific cultural needs and infrequent blooming characteristics, many orchids are not as widely distributed as other garden plants and primarily remain as collector's favourites. Those that are popular in gardens include the following: -

Arachnis Maggie Oei

Spider / Scorpion orchids

This is a strong-stemmed semi terrestrial monopodial orchid adapted to open and lightly shaded humid situations prevailing along the west coast.

They climb vigorously on to live and dead supports. They are free flowering and produce axillary many flowered inflorescences that often have short basal branches. The tepals that are held apart like spider's legs are coloured greenish yellow to creamy white and marked with pink or pale brown to dark brown blotches and bands and the lip is creamy white to mauve in colour. The different flower colour variants are named Red Ribbon, Yellow Ribbon, Pink Ribbon and Brown Maggie. These have cut flower value.

Arundina graminifolia (Don) Hoch.
Syn. *Arundina bambusifolia* Lindl.

Bamboo orchid

These are terrestrial to semi-epiphytic sympodials that become bushy and form large clumps. The leaves are linear-lanceolate, grass-like and held on pale yellow cane-like shoots resembling bamboo shoots. The flowers are fragrant and coloured mauve-pink with a dark pink lip, with prominent bracts and are produced terminally on the shoots.

Cattleya skinneri Bat.

Cattleya

These are pseudobulbous epiphytes. The smooth green pseudobulb terminates in one or two fleshy leaves. The few flowered inflorescences are also borne terminally in the axils of these leaves, subtended by a spathe-like bract. The perianth lobes are mauve-pink in colour. Plants are grown in pots or on suitable tree-supports.

Dendrobium Sw.

Dendrobium orchids

These are sympodial epiphytes that have fleshy shoots and leaves with well-marked veins and thin roots. They are grown in pots placed a meter or so above the ground or hung overhead or are grown on tree supports. A large number of hybrids have been evolved in Thailand, Singapore and Malaysia in crosses involving more than 80 species and their varieties. Their flower colours range from greenish white to dark purple and golden yellow, with contrasting lip colours and striped tepals. In gardens of the west coast of India, very old varieties to the popular cut flower varieties of today are grown. These include *Dendrobium* Caesar White and Caesar Pink, *Dendrobium*

Philippica, *Dendrobium* Madam Pompadour, *Dendrobium* Jacquelyn Concert, *Dendrobium* Sonia –17, 28, Sonia Super Queen and Sonia Bom-Jo, *Dendrobium* Sariffa Fatimah, *Dendrobium* Tongchai Gold, *Dendrobium* Kasem Gold, *Dendrobium* Burana Fancy, *Dendrobium* Ekapol Panda, *Dendrobium* Candy Stripe, *Dendrobium* Nagoya Pink, *Dendrobium* Rinnapa, *Dendrobium* Sabine, *Dendrobium* New Wanee, *Dendrobium* Sakura, *Dendrobium* Emma White, *Dendrobium* Thailand White and *Dendrobium* Snow White.

Dendrobium crumenatum

Malaysian Pigeon orchid

These are epiphytes having shoots that are thickened up to the lower 4th or 5th node only. Inflorescences are produced terminally and carry small fragrant white to pink tinged flowers that open out in response to a sudden drop in temperature following a summer shower. The flowers are ephemeral. The species is widely distributed in South East Asia

Epidendrum ibaguense
Syn. *Epidendrum radicans*

Holy cross orchid

Plants are erect to decumbent epiphytes that have thin shoots rooting readily from the nodes. The leaves are small and ovate. The flowers are produced in compact terminal racemes. The flowers are yellow, orange or red with faint darker markings on the lobes of the lip, which is shaped like a cross.

Oncidium Sw.

Dancing Girl orchids

These are pseudo-bulbous epiphytes with ovate thick and fleshy or lanceolate thin and leathery leaves. The flowers have perianth lobes coloured white or yellow with brown, maroon or red spots and blotches. The lobed lip is spread outward resembling a frilled skirt. *Oncidium* Goldiana is commonly grown in Kerala.

Peristeria elata

Dove orchid

Plants are terrestrial sympodials with hard smooth green pseudobulbs that carry plicate leaves terminally. The inflorescences are produced from the basal nodes of the pseudobulb. They carry 12 to 25 waxy fragrant white flowers that have cupped perianth lobes. The winged lip with its lateral lobes appears dove-like.

Phalaenopsis Blume

Moth orchids

These are monopodial epiphytes with condensed stems, adapted to grow well and flower at subtropical climates. The leaves are fleshy and the shoots are slow growing. The inflorescence axis is zigzag in shape. The lateral petal lobes have rounded edges that droop giving them a shape like the outspread wings of a moth at rest. Inter generic hybrids with *Doritis* are also grown.

Renanthera Lour.

Red Vandas

These are spectacular epiphytic monopodials native to the North Eastern hill zone of India up to Burma. Erect to clambering in habit, they have flowers coloured predominantly red with overtones of yellow and faint darker red markings. *Renanthera coccinea, Renanthera imschootiana* and *Renanthera storiei* collected from native habitats have been distributed throughout the world and used in breeding programmes to impart the relatively rare red colour to orchids.

Rhyncostylis retusa

Fox tail orchid

Plants are slow growing monopodial epiphytes with strap-like leaves that are held distichously. They are found naturalized atop shade and avenue trees of gardens in large and loose clumps. Inflorescences are produced after the summer showers of March-April. Flowers are small, pinkish white and compactly arranged on a drooping peduncle.

Spathoglottis plicata

Ground orchid, Spathoglottis

Plants are terrestrials with ovoid pseudobulbs and plicate leaves. Inflorescences arise from basal leaf axils. The long peduncles carry flowers that are yellow, red, orange or light to dark purple. *Spathoglottis aurea* has white flowers.

Vanda R. Br.

Vandas

These are monopodials that are grown in gardens as semi terrestrials. The leaves are strap-like to completely terete. The flowers are held uniformly around the peduncle. Old varieties that are garden favourites include the terete leaved *Vanda* Miss Joaquim and *Vanda* Diana. Species used extensively for breeding include *Vanda caerulea, Vanda tesselata Vanda teres, Vanda tricolor and Vanda sanderiana.*

Aranda

Aranda orchids

These are bigeneric hybrids of *Arachnis* and *Vanda* orchids. A large number of them have been evolved and are commercially grown as cut flowers. They are intermediate in inflorescence length when compared to parental genera and in bloom characters combine the better traits of both. A number of introduced varieties such as Aranda Chark Kuan, *Aranda* Peter Evert, *Aranda* James Storiei, *Aranda* Nancy, *Aranda* Iskander, *Aranda* Christine, *Aranda* Tay Swee Keng, *Aranda* Eric Mc Kie *Aranda* Lucy Laycock and *Aranda* Norah Alsagoff are grown.

Aranthera

Red Vanda hybrids, Arantheras

These are bi-generic hybrids of *Renanthera* and *Arachnis* species. Their flowers combine the red shades of the former with the tepal size and shape of the latter. They are commercially grown cut flowers. Important cultivars are *Aranthera* Anne Black, *Aranthera* James Storiei and *Aranthera* Mohammed Haniff.

Mokara

Mokaras

These are trigeneric hybrids involving *Vanda, Ascocenda* and *Arachnis.* They are slow growing, floriferous and combine a range of flower colours in relatively short inflorescences. Widely adapted cultivars include *Mokara* Chark Kuan series with pink orange-yellow and white spotted flowers.

COOL SEASON PERENNIAL FLOWERING HERBS

These are adapted to grow and flower profusely in temperate latitudes. **Cool season perennial flowering herbs** popularly grown in gardens include the following: -

Acantholimon androsaccum Plumbaginaceae

Acantholimon, Prickly Thrift

Plants are low growing evergreens suited for dry and exposed situations. The flowers are purple- mauve.

Achillea filipendulina Asteraceae
syn. *Achillea eupatorium*

Achillea, Milfoil

Plants are bushy and attain heights of up to 1.5m. The leaves are pinnate, toothed and aromatic. Flowers are yellow and borne in dense heads.

Achillea millefolium Asteraceae

Milfoil, Yarrow

Plants grow to a height of up to 1m. Flowers are red or purple. Propagated by division of clumps.

Ajuga reptans Lamiaceae

Ajuga, Blue bugle, Bugle herb, Bugleweed, Common bugle, Carpetweed

Plants are low growing and spreading evergreens. The leaves are coloured bronze-red. The flowers are pink or blue in colour. They are suited for partially shaded areas of gardens.

Ammobium alatum Asteraceae

Winged Everlasting flower

Plants have ridged stems with soft silvery leaves. Flowers are white with yellow centers and arise in clusters. They resemble *Helichrysum* flowers.

Anchusa angustissima Boraginaceae

Anchusa, Perennial Bugloss

Plants are low growing erect evergreens that thrive in open situations. The flowers are blue.

Anthemis tinctoria Asteraceae

Anthemis, Yellow Chamomile flower

Plants are erect herbs attaining heights of 1.2m. The leaves are finely cut. Flowers are yellow and are produced in summer. *Anthemis nobilis* is the Roman Chamomile.

Aquilegia vulgaris Ranunculaceae

Columbine

Plants are erect clump-forming herbs attaining heights of over one meter. The leaves are trifoliate. Flowers are single or double, having a long spur. They are produced during spring and summer seasons. Plants are propagated by seeds or by division of clumps.

Aster novi-beigii Asteraceae

New York Aster

Plants are clump-forming herbs attaining heights of up to 1.2m. They are suited for semi-shaded and open areas. The flowers are white, pink, purple or blue and are producing in autumn.

Bellis perennis Asteraceae

Daisy, English daisy

Plants are dwarf in stature and have strap-like leaves growing in a tuft. They can be grown as annuals. The flowers are single or double, produced on long stalks with petals that are broad and flat or quilled. They are coloured white, pink or red. Propagation is by seeds or suckers.

Bergenia cordifolia Saxifragaceae

Elephant ear

These are cold-hardy spring-flowering perennials native to Siberia. The leaves are heart shaped and bronze tinged. The flowers are rose-pink. Plants are propagated by seeds or by division of roots. *Bergenia ligulata* found in Nepal has white or pink flowers.

Browallia speciosa Solanaceae

Amethyst flower, Browallia, Bush Violet, Sapphire flower, Star flower

Plants attain heights of up to 0.5m or more. The flowers are bright blue. Plants are adapted to partially shaded situations and are propagated by seeds and cuttings.

Campanula persicifolia Campanulaceae

Tussock Bellflower

Plants are clump-forming, erect evergreens attaining a height of about 1m. Flower colours range from white to violet or blue-mauve. Plants are propagated by division of clumps. *Campanula carpatica* is also a perennial grown in gardens.

Cynoglossum amabile Boraginaceae

Hound's tongue, Chinese Forget-me-not

Plants are hardy perennials that are often grown as annuals in gardens. Very small, blue or white scented flowers are produced in clusters. They are ideal for rockeries and for partially shaded situations.

Chrysanthemum maximum Asteraceae

Shasta Daisy / Moon Daisy

Plants are hardy perennials attaining heights of up to 0.5m. they are summer flowering plants. The flowers have white rays and yellow disc florets

Delphinium belladonna Ranunculaceae
Syn. *Delphinium hybridon / Delphinium hybrida*

Perennial Delphinium

Plants are short in stature, attaining heights of up to 1m. Flowers are in various shades of blue. They are ideal for massing.

Dianthus barbatus Caryophyllaceae

Sweet William

Plants are dwarf in stature and are biennials that have pink to red flowers having faint darker markings.

Dianthus plumarius Caryophyllaceae

The Cottage Pink

Plants are perennials considered to be hybrids of Pinks and Sweet Williams.

Digitalis grandiflora Scorphulariaceae

Yellow Foxglove

Plants are erect evergreens attaining a height of up to 1m suited for partially shaded situations. The flowers are yellow and are produced in summer. They can be grown as annuals.

Gaillardia pulchella Asteraceae

Blanket flower

Plants are hardy semi-perennial herbs, which thrive in mild temperate regions and in the humid tropics. They are ideal bedding plants and can be grown as annuals. They are propagated by seeds. Flowers are held on long stalks above the canopy level. Their colours have a blend of yellow, cream, maroon and orange.

Gazania pavonina Asteraceae

Peacock Gazania

Plants are hardy dwarf growing perennials having white, pink, yellow or orange flowers that are daisy like. *Gazania splendens* called the treasure flower is also grown. A number of hybrids of these two with a range of flower colours are grown in temperate gardens.

Geranium macrorrhizum Geraniaceae

Cranesbill

Plants are low growing prostrate evergreens having variegated and scented leaves. They are summer flowering and suited for exposed areas. The flowers are white, red or pink.

Gerbera jamesonii Asteraceae

Transvaal daisy, Barberton daisy, Gerbera

Plants are low growing herbs having obovate and deeply lobed leaves held in a rosette on a condensed stem. The flowers are single to double up to 10cm in diameter and are borne on erect thick stalks. The rays are coloured in pastel shades or brightly and the disc florets usually have a contrasting colour. Plants are grown in mild temperate and sub tropical situations for cut flowers.

Geum bortsii Rosaceae

Avens, Geum

Plants are shade loving perennial herbs having reddish coloured rose-like flowers. Flowering is in late summer.

Gypsophila cerastioides Caryophyllaceae

Himalayan Baby's Breath, Rock Gypsophila

Plants are evergreen herbs that are dwarf in stature with a spreading habit. The leaves are silvery green in colour. Flowers are mostly white or pink. Mauve flowered varieties are also grown. *Gypsophila paniculata* has single or double flowers. *Gypsophila nepens* has pink flowers and a trailing habit.

Heuchera sanguinea Saxifragraceae

Coral bells, Crimson bells

Plants are evergreens attaining heights below 0.5m. The leaves are rounded and lobed. Flowers are small bell shaped, red and are borne on long stalks in profusion from late summer to early autumn. Plants are propagated by seeds and by separation of clumps.

Hunnemannia fumariifolia Papaveraceae
Syn. *Hunnemannia fumariaefolia*

Mexican Tulip-Poppy

Plants are perennials attaining heights below one meter. The flowers are yellow and up to 7cm in diameter. Plants flower late in summer. Propagation is by seeds and separation of suckers.

Linum perenne Linaceae

Perennial flax

Plants are blue or white flowered. Propagation is by seeds or cuttings.

Lupinus polyphyllus Fabaceae

Lupin

Plants are erect clump-forming perennial herbs suited for open situations. The flowers have a wide range of colours like white, pink, purple and yellow.

Oenothera acaulis Onagraceae

Perennial Evening Primrose

Plants are spreading herbs adapted to semi-shaded or sheltered situations. Flowers are yellow or pink.

Paeonia officinalis Ranunculaceae

Peony

Plants are clump forming, summer flowering erect perennials attaining heights of about a meter. The leaves are ornamental. The flowers are creamy white to red or pink. Propagation is by division of clumps.

Pelargonium hortorum Geraniaceae
Syn. *Pelargonium domesticum*

Pelargonium, Geranium

Plants are succulent perennial herbs adapted to temperate situations popularly grown as bedding plants or in containers. Varieties grown in gardens are classified based on leaf characters into Zonal / Zoned / Horse-shoe Pelargoniums that have concentric horse shoe like rings on leaves, 'Scented-leaved Pelargoniums' having rose scented or nutmeg scented leaves and 'Ivy-leaved Pelargoniums' having deeply toothed leaves. Flowers are produced in clusters and are generally white to pink, salmon, vermilion, crimson or purple.

Penstemon barbatus Scorphulariaceae

Beard Tongue

Plants are hardy erect clump forming evergreens adapted to partially lighted situations. The tubular flowers are yellow, red or pink. Flowering takes place in summer.

Phlox subulata Polemoniaceae

Creeping Phlox, Moss Pink, Moss Phlox, Perennial Phlox, Thrift

Plants are dwarf evergreens suited for growing in containers and in exposed areas. They are spring and summer flowering plants. The flowers are white, red or pink to mauve.

Primula malaccoides Primulaceae

Baby Primrose, Fairy Primrose

Plants are short in stature. The leaves are borne in a basal rosette. The flowers are single or double and coloured in varying shades of red or white. *Primula denticulata* having mauve flowers and *Primula rosea,* adapted to moist situations are also grown in gardens.

Pulmonaria rubra Boraginaceae
Red Lungwort, Pulmonaria

Plants are dwarf evergreens with a spreading habit. They are spring and summer flowering. The flowers are red. Plants are suited for growing in shaded situations and in containers or as a ground cover.

Pyrethrum roseum Asteraceae

Pyrethrum

Plants are erect clump forming herbs attaining heights of up to 1m. They have ornamental foliage. Plants are summer flowering in nature. The flowers are white, pink or red in colour.

Rudbeckia hirta Asteraceae
Syn. *Rudbeckia fulgida*

Black-eyed Susan

Plants are erect clump-forming biennial herbs. They flower late in summer to autumn. The flowers are yellow, red or bicoloured.

Solidago Canadensis Asteraceae

Golden rod

Plants are clump-forming low growing perennials, attaining a maximum height below 1m. They are suited for semi-shaded and well-lighted situations. The inflorescences produced terminally are feathery and have minute flowers. The inflorescences are popularly used decoratively. Plants are propagated by suckers that are produced in profusion at the base of flowered shoots. These can also be grown in mild temperate and sub-tropical situations.

Streptocarpus hybridus Gesneriaceae

Cape Primrose

Plants are stemless perennials that are prostrate in habit. The leaves are formed in a rosette. Flowers in a range of colours are borne on erect scapes. Plants thrive equally well in open and sheltered situations. Propagation is by seeds and leaf cuttings.

Strobilanthus atropurpureus Acanthaceae

Persian Shield, Strobilanthus

Plants are clump forming erect herbs attaining heights of up to 1m. They are autumn flowering. The flowers are tubular and purple to mauve.

Verbena hybrida Verbenaceae

Hybrid Verbena, Florist's Verbena

Plants are short in stature and are slender stemmed semi perennials with a spreading habit that are grown as annuals in temperate regions. The flowers are small and white, pink, purple or mauve and are held in small clusters.

Veronica gentianoides Scorphulariaceae

Veronica

Plants are low growing herbs that are spring and summer flowering in nature. The leaves are ornamental. The flowers are blue.

Viola cornuta Violaceae

Pansy

Plants are dwarf spreading semi perennials adapted to partially shaded situations. They are spring and summer flowering in nature. The flowers are fragrant and coloured predominantly in shades of purple-violet.

PERENNIAL FOLIAGE HERBS

Foliage Herbs of tropical gardens include monocots and dicots that are primarily grown for their decorative foliage. Most of them are adapted to shaded and partially shaded situations and a few can thrive in open sunny situations of the tropics.

MONOCOTYLEDONOUS PERENNIAL FOLIAGE HERBS

These are non-woody perennial monocots adapted to shade and are grown for their ornamental foliage. They include the following clump-forming herbs and creeping or climbing semi-epiphytes. Popular **monocotyledonous perennial foliage herbs** of Indian gardens include the following: -

Aglaonema Schott Araceae

Aglonema, Chinese Evergreen

These are evergreens having stems that gradually acquire a reclining habit. New shoots are produced from the basal nodes giving plants the appearance of clumps. The leaves are variegated and ovate or lanceolate. The base of the petiole encircles the stem like a sheath and leaves a translucent layer after the leaves are shed. This distinguishes them from Dieffenbachias. The important species into which cultivated varieties are grouped are: -

Aglaonema commutatum Schott

Silver Queen Aglonema

Plants have leathery dark green leaves with silvery white markings. They are ideal for massing in pots indoors and in partially lighted areas.

Aglaonema costatum

Fox's Aglonema

Plants are short in stature and have broadly ovate leaves with yellow to white spots on the lamina and a broad band through the midrib

Anthurium Schott Araceae

Tail flowers

Plants of this genus are shade loving creeping or climbing herbs with semi-epiphytic adaptations. Apart from the species commercially grown for flowers – see under **Warm-season Perennial Flowering Herbs** - the following species that have unusually attractive leaves are popularly grown as **Foliage Anthuriums**: -

Anthurium clarinervium

White-veined Anthurium

These are succulent stemmed and slow growing epiphytes attaining a creeping or climbing habit with age. The leaves are ovate to chordate and up to 0.5m in length and width. They are velvety dark green in colour with well-marked pale greenish white veins. Their reticulate venation is spectacular and uncommon in monocots. The basal lobes are absent or distinct and may be overlapping or not.

Anthurium crystallinum

Crystal Anthurium

Plants are more robust than the above species with larger leaves that are up to 1m in length and width. The basal lobes are overlapping or not. They are suited for display in large indoor areas.

Anthurium veitchi Mast.

King Anthurium

Plants have elongated pendant leaves with well-marked and depressed parallel veins. The leaves are nearly 10cm wide and attain a length of more than 1m.

Anthurium bakeri Hook.

Ribbon-leaf Anthurium

Plants are unique semi-epiphytes having strap like pendant leaves of more than 90cm length and 5cm width.

Asparagus densiflorus Liliaceae
Syn. *Asparagus sprengeri*

Fox-tail Asparagus, Sprenger Asparagus

Plants have bushy columnar shoots that recline with age. They are ideal for pot display

Asparagus sprengeri cv. Meyerii has more compact shoots.

Asparagus officinalis Linn.

Asparagus racemosus Willd.

Garden Asparagus (Eng.), Sathavari (Sans.)

Plants are shade-loving climbers having bright green phylloclads that are sickle shaped and shiny which make them excellent foliage plants. The stems are smooth, shiny and thorny. Cut foliage is used for decorative purposes.

Asparagus plumosus Baker
Syn. *Asparagus setaceous*

Asparagus fern, Evergreen fern

This species has small and finely divided phylloclads held on a horizontal plane. They are extensively used as decorative foliage. *Asparagus plumosus* var. *nanus* is dwarf growing and bushy in habit and has short internodes and small sized leaves.

Asparagus pyramidalis

Ball asparagus

Plants have erect shoots that are sparsely thorny. The phylloclads are bright green in colour and clustered towards the tips of shoots and appear ball-like. The branched upright shoots have an overall pyramidal outline.

Asparagus retrofractus Forsk.
Syn. *Asparagus macowanii*

Ming Fern, Pom-pom Asparagus, Zig-zag shrub

Plants are clump forming and dwarf in stature. The shoots are thin and recline with age. The phylloclads are sickle shaped and closely arranged on the shoots. Plants are ideal for pot display and are propagated by shoots having fleshy roots obtained by separation of the clumps.

Calathea G. F. W. Mey Marantaceae

Calathea

Plants are shade-adapted clump forming perennials with short thickened stems. The leaves are broadly ovate and the petioles are thin and rounded along the upper half and the lower half is broad and sheath-like. The leaves are attractively marked with white or light green colour. The inflorescences are ginger-like. Plants are indigenous to S. America. The important species grown in our gardens are the following: -

Calathea ornata Koern.

Shadow plant

Plants are robust attaining heights of up to 1m. The leaves have pink and white lines on the upper surface.

Calathea zebrina Lindl.

Zebra plant

Plants are dwarf in stature. The leaves are ovate-lanceolate with broad and green markings.

These are the most striking among the commonly grown calatheas and are sensitive to high temperatures and low humidity in the environment and low moisture content in the growing medium.

Chlorophytum comosum Baker Liliaceae

Spider plant

Plants are clump-forming perennials. The leaves are strap like with pointed tips. They have a creamy yellow band running along the middle of the lamina. The inflorescences carry small white flowers. After the flowers are shed, the inflorescences produce axillary shoots, which hang down like spiders when plants are pot grown or hung overhead in hanging containers. Natural vegetative propagation is achieved when the inflorescence axis with the hanging shoots having root initials reach the ground.

Chlorophytum indicum

Chlorophytum

Plants are similar in habit to the above species. The leaves are leathery and longer. The flowers are pale yellow in colour. However the production of shoots on inflorescences is not observed.

Cyperus alternifolius Cyperaceae

Umbrella sedge, Umbrella plant, Umbrella Palm sedge

Plants are popular ornamentals of gardens. The shoots and leaves are dark green and glossy. The leaves are held in a spreading spiral towards the tips of erect shoots like an umbrella.

Dieffenbachia Schott Araceae

Dumb cane

Plants have succulent unbranched stems that have cane like markings. The leaves are elliptic-lanceolate with white, creamy white or light green mottling or marbling on the lamina. They are extensively grown as indoor and outdoor foliage plants. Cultivars differ chiefly in their leaf patterns. The plant sap is rich in oxalates. Hence ingesting plant parts by chewing cause irritation and swelling of the upper respiratory tract rendering victims temporarily speechless. This has resulted in the plants being called Dumb canes. The important species into which varieties are grouped include *Dieffenbachia picta* in which, the leaf margins and midribs are dark green and the lamina is variously blotched with ivory white lines and blotches, *Dieffenbachia hoffmanii* having mottled green leaves with marbling of a darker green colour and *Dieffenbachia maculata* having light green leaves lined along the margins and mid rib with dark green.

Epipremnum aureum Schott Araceae
syn. *Scindapsus aureus* Engl. / *Pothos aureus*

Variegated Pothos, Devil's Ivy, Hunters Robe

Plants are evergreen semi-epiphytic climbers. They thrive in shaded and partially lighted situations. The leaves are characteristically variegated or marbled. Plants undergo a juvenile-adult change in the shape of the leaves. The juvenile leaves are entire and chordate while the adults have broadly ovate leaves with irregularly cut margins.

Homalomena rubescens Kunth. Araceae

Borneo Purple plant, Homalomena

Plants are evergreen clump forming herbs. The leaves are chordate and attain a length of up to 35cm. Under dense shade, the leaves remain predominantly green while under partially lighted conditions, they attain a reddish colour along the veins. The petiole and the under surface of the leaves are reddish maroon in colour.

Maranta Plum. ex Linn. Marantaceae

Maranta

Plants are moisture loving and shade adapted succulent herbs. They are sensitive to high temperatures and low moisture levels in the growing medium, showing in rolling of the lamina under both situations. The leaves have a velvety surface texture. They are ovate and with intricate markings of dark green, pale green and pink on the lamina. A number of species have been differentiated such as *Maranta leuconeura* E. Morr. syn. *Maranta kerchoviana* with regular dark green patches on either sides of the mid rib and *Maranta massangeana* having more colourful patches and lines on the lamina. *Maranta variegata* has creamy-yellow variegations on the leaves.

Monstera Schott Araceae

Swiss cheese plant

These are shade adapted thick-stemmed semi-epiphytic climbers having aerial roots that hang down. Leaves are broadly ovate with characteristic perforations. Two species grown in gardens are *Monstera pertusa* Schott syn. *Monstera adansonii,* which are robust climbers and have small rounded to elongated perforations on the leaves and *Monstera deliciosa* Liebm., that is similar in habit to the above species. In these the leaves have perforations of various sizes and in mature plants the lamina has deeply cut margins.

Ophiopogon intermedius D. Don Ruscaceae

Lily Turf, Mando grass, Ribbon grass

Plants are evergreen herbs with a bushy habit and are adapted to partial shade. The leaves are dark green, narrowly strap-like and attaining a length of more than 40cm and a width of about 1cm and arch gracefully. In a variegated selection, the lamina has longitudinal white lines.

Philodendron Schott Araceae

Elephant-ear plants, Philodendrons

As the name implies, these are epiphytic creepers or climbers adapted to shade. They are native to tropical America and a number of species and varieties have introduced into India. They are excellent indoor plants. Leaves are entire to deeply lobed. Important species grown in Indian gardens include *Philodendron gloriosum* Andre in which the stems are thin and relatively tender with small dark green heart shaped leaves having snow-white veins. The leaf margins are thinly pink-edged. The young leaves have a silky surface texture.

In *Philodendron lacinatum* Engl., the leaves are large, ovate-chordate with finely cut and wavy margins. They droop attractively. In *Philodendron scandens* C. Koch & Sello., the most popular among the species grown and called the Heart-leaf Philodendron, plants are thick-stemmed robust creepers or climbers.

The leaves are ovate to heart-shaped, entire, leathery and shiny. They are coloured light green to dark maroon red. The popularly grown cultivars are 'Royal King' having light green coloured leaves,'Emerald Duke' having dark olive green leaves, 'Emerald Red' having dark green leaves with red petioles and midrib and 'Red Prince' having entirely dark maroon coloured leaves.

Philodendron squamiferum Poepp. & Endl. are thin-stemmed climbers. In young plants the leaves are entire and in older plants they are three lobed. The petiole and young shoots are entirely covered with soft and curved reddish green bristles. The older shoots are sparsely bristled. *Philodendron bipinnatifidum* Schott are robust slow growing creepers. The leaves are pinnately lobed with wavy margins. *Philodendron goeldi* are unique plants with pinnate leaves having a curved rachis. The leaflets are of unequal length and attached to the outside of the curved rachis in a radiating manner.

Rhoeo spathacea Commelinaceae
Syn. *Rhoeo discolor*

Moses-in-a-boat, Rheo

Plants have thick leaves coloured dark violet on the lower surface. The axils carry white flowers in boat-shaped bracts.

Syngonium podophyllum Schott Araceae

Arrowhead vine

Plants are vigorous bushy creepers. The leaves are green with white variegations along the veins. The lamina is arrowhead like and separated into three lobes. The juvenile leaves are simple. Plants are very popularly grown indoors.

Tacca integrifolia Ker-Gawl. Taccaceae

Cat's Whiskers, Bat flower

Plants are clump-forming shade loving perennials having fleshy roots. The leaves are glossy dark green, broadly ovate-lanceolate. The flowers are produced on an erect scape. They have two pairs of small and large bracts coloured purplish black. The bracteoles are thread-like and are held arching on either side like whiskers.

Tradescantia blossfeldiana Commelinaceae

Spiderwort

Plants are trailing herbs with succulent stems. Leaves are small ovate-lanceolate and variegated with silvery white and purple lines. They are quick growing ground covers suited for shaded situations.

Tradescantia pallida Commelinaceae
syn. *Setcresea purpurea*

Purple heart, Purple queen

These are shade loving succulent creepers and are entirely purple coloured. The flowers are pink. They are ideal for providing colour in shaded situations.

Tradescantia zebrina.
Syn. Zebrina pendula

Inch plant, Wandering Jew

These are succulent creepers with bright green leaves that have two silvery bands on the lamina. The underside of leaves is purplish green.

DICOTYLEDONOUS PERENNIAL FOLIAGE HERBS

This small group has a few sun loving plants such as *Alternanthera,* Foliage Amaranths, *Oscimum* and *Plectracanthus* and a number of shade adapted ornamentals like *Begonia, Peperomia* and *Pilea.*

The **dicotyledonous perennial foliage herbs** commonly grown in Indian gardens include the following: -

Alternanthera Forsk.
Alternanthera versicolor Hort. ex Regel. Amaranthaceae

Poocheera

Plants are dwarf herbs grown extensively in gardens as carpet bedding plants and for pattern planting. Cultivated varieties have small green or pink leaves with variegations of paler shades.

Alteranantheral brasiliana Amaranthaceae

Purple Chaff flower / White Buttons

Plants are perennial herbs attaining a height of 0.5m under partially shaded situations. In the open, they are low growing, attaining height of less than 30 cm. Under full sunlight plants are entirely maroon-red in colour with small white button-like flower heads. The leaves are ovate and upto 5cm in length.

Amaranthus bicolor Amaranthaceae

Foliage Amaranth

Plants are annual herbs grown in gardens for their attractive foliage. They attain heights of up to 1.2m. The foliage becomes maroon tinged and the top becomes crimson- red. In *Amaranthus tricolor* Linn. the foliage attains 3 different colours. Cultivar 'Splendens' has scarlet terminal leaves and 'Rubra' has bright red leaves.

Amaranthus salicifolius Hort. Vietch ex Gard. Amaranthaceae

Weeping Amaranth

The plants have foliage that characteristically droops downwards. The leaves have wavy margins and are held close to the main shoot. They are bright orange to dark red in colour.

Begonia (Tourn.) Linn. Begoniaceae

Begonia

Plants are perennial herbs attaining heights of up to 1m, with succulent plant parts and attractive leaves. They are shade loving and adapted to humid situations. Some are found naturalized as lithophytes. A number of species are grown in gardens. Those that are popular in tropical gardens are *Begonia rex* Putz. and *Begonia semperflorens* Hook. *Begonia rex* cultivars are called 'Rex Begonias'. They are characterized by the intricate patterns and unusual colour tones on the thick coarse textured leaves. The leaves are large and showy while the unisexual flowers are inconspicuous. Rex-Begonias are indigenous to N.E. India. They are propagated by leaf cuttings. *Begonia semperflorens* cultivars are referred to as '**flowering begonias**'. Their flowers are conspicuous and attractive.

Species and cultivars of *Begonia* are also classified based on their growth habit and leaf characteristics as belonging to the following groups.

1. Tuberous- rooted or Rhizomatous begonias

They are plants that have thick rhizomatous stems that are decumbent or prostrate. Axillary shoots are produced from these stems giving plants a spreading habit.

2. Fibrous- rooted begonias

These produce a tuft of leaves from a very condensed stem found below the surface of the growing medium. Axillary shoots are produced basally from the stem resulting in clumps.

3. Cane-stemmed begonias

The rhizomatous stems of these are thinner and cane-like.

4. Angel-wing begonias

These have obliquely ovate leaves shaped like wings

5. Lotus-leaved begonias

These have sub-orbicular leaves held on long petioles. The petioles are attached to the basal end of the lamina (peltate) and the leaf margins are serrulate giving them an overall resemblance to the leaves of lotuses and lilies. The species *Begonia nelumbiifolia* Cham & Schlecht, indigenous to central and South America is characterized by these leaves.

Fittonia verschafeltii E. Coem. Acanthaceae
Syn. *Fittonia argyroneura* E. Coem.

Mosaic plant, Nerve plant

Plants as shade adapted perennial herbs having colourful foliage. The leaves are ovate, 5 to 7cm long and well marked with reticulate venation. The veins are pink or white in colour. Plants tend to spread slowly like a ground cover in moist and cool situations when grown in a well-drained humus-rich medium.

Hemigraphis colorata var. *exotica* Acanthaceae

Purple Waffle plant

These are low growing bushy plants adapted to shaded and partially shaded situations. Unlike the parental species, they do not creep readily. The foliage is green on the upper surface and purple beneath. The leaves are ovate and the lamina is undulate between the veins, giving them the appearance of being slightly crinkled.

Hypoestes Soland. Ex R. Br. Acanthaceae
Hypoestes sanguinolenta Hook.
Syn. *Hypoestes phyllostachya*

Polka-dot plant, Freckle-Face plant

Plants are evergreen herbs with succulent stems adapted to partially shaded situations. The dark green leaves are up to 5 cm in length and spotted with pink spots of various sizes. Cultivars having smaller leaves that are densely spotted with magenta-pink or white spots have been evolved, belonging to the 'Persuasion' series.

Iresine herbstii Hook. Amaranthaceae

Blood leaf

Plants are evergreen herbs to sub-shrubs having attractively reddish green coloured shoots and leaves. The leaves are kidney shaped with indented tips and coloured green with red veins or reddish purple with bold markings along the veins. Plants are semi perennial in habit and can be massed in pots for attractive colour in indoor areas and shaded situations.

Iresine diffusa Humb.& Bonpl. Ex Willd. Amaranthaceae
Syn. *Iresine lindenii*

Blood leaf

Plants are similar in habit to *Iresine herbstii*. The leaves are ovate, 5 to 7 cms in length with acuminate tips and are blood-red in colour with white or pale pink veins.

Orthosiphon aristatus Miq. Lamiaceae
Syn. *Clerodendron spicatum* Thunb., *Orthisoiphon stamineus* Hort.

Cat's Whiskers

Plants are perennial sub shrubs adapted to partial shade. The leaves are ovate with serrate margins. The white flowers are produced on terminal spikes. The floral appendages that arch downwards have a resemblance to whiskers.

Oscimum sanctum Lamiaceae

Krishna Thulasi, Manjari, Sacred Tulsi, Sacred Basil

Plants are semi perennial, erect, pubescent herbs. The aromatic green leaves become purple tinged in sunny situations. Minute purplish aromatic flowers are held in erect terminal racemes. Plants are revered in India. The leaves and flowers have varying medicinal uses and are extensively used in home remedies for common cold and fever, as insect baits, for hair adornment and are also made into garlands.

Pedillanthus tithymaloides Poit. Euphorbiaceae
Syn. *Euphorbia tithymaloides, Pedillanthus canaliculatus* Sweet, *Pedillanthus carinatus* Spreng.

Devil's backbone, Redbird cactus, Slipper spurge

Plants are succulent herbs adapted to partial shade and attain heights of up to 1m. The leaves are thick, ovate and tinged pale green to white and pink. The stems are darker green in colour and are zig-zag in shape. In *Pedillanthus tithymaloides* var. *nanus,* the shoots are dark green, thinner and shorter with less branching and having distichously arranged leaves. These are suited for edging.

Peperomia argyreia E. Morr. Piperaceae
Syn. *Peperomia sandersii* C.DC.

Peperomia

Plants are succulent herbs adapted to shady and humid locations. A large number of selections with thick ovate leaves having rounded to broadly acute tips with green, maroon-red and creamy white colouration or variegations on the lamina are grown in gardens and indoor areas. They are excellent for table-top display.

Pilea Lindl. Urticaeae

Pilea cadieri

Aluminium plant

Plants are shade-adapted herbs with attractive foliage. The leaves are dark green with regular shiny white patches on the lamina, which give them the appearance of being painted on. The flowers are creamy white.

Pilea microphylla Liemb.
Syn. *Pilea muscosa* Lindl.

Pilea

These are dwarf and succulent shade and moisture loving herbs that are frequently found in moss covered areas. The leaves are minutely lobed and held spread out.

Pilea repens Liemb. Urticaceae
Syn. *Pilea nummulariaefolia* Wedd.

Creeping Charley

Plants have small rounded succulent green leaves and red tinged shoots. They tend to creep and spread in shady humid situations. They are ideal for growing in hanging pots. *Pilea serpyllacea* has larger leaves.

Plectranthus scutellarioides Lamiaceae
Syn. *Coleus blumei, Coleus verschafeltii, Solenostemon scutellarioides*

Coleus, Painted nettle

Plants are semi-perennials herbs having succulent, angled stems and leaves that are coloured maroon-red, yellow and green. The leaf margins are attractively serrate. Plants are moisture loving and leaves become brightly coloured under well-lighted and partially shaded situations. Plants attain heights of up to 1m and are grown like annuals in the humid tropics of India. Branching and leaf growth are retarded after the production of terminal inflorescences on the shoots. The shoots carry root-initials that enable plants to root readily.

contd. colour plates

Angelonia salicariaefolia

Angelonia salicariaefolia 'Alba'

Begonia rex

Begonia semperflorens

Impatiens hawkerii

Impatiens hawkerii 'Busy Lizzie'

Anthurium andreanum 'Tropical'

Centratherum intermedium

Chrysothemis pulchella

Gerbera jamesonii

Spathiphyllum cannaefolium

Tacca integrifolia

Impatiens balsamina

Portulaca grandiflora

Salvia splendens

Tithonia rotundifolia

Torenia fournieri

Zinnia elegans

Aster amellus

Cosmos bipinnatus

Dahlia variabilis

Dendranthema grandiflorum

Petunia hybrida

Tagetes patula

Alpinia variegata

Calathea ornata

Coleus blumei

Costus speciosus 'Variegata'

Iresine herbstii

Peperomia argyreia

Aranthera 'Anne Black'

Doritaenopsis

Mokara 'Chark Kuan Pink'

Mokara 'Singapore Red'

Phaleanopsis

Rhyncostylis retusa

Dendrobium 'Caesar Pink'

Spathoglottis plicata

Dendrobium 'Sonia Super Queen'

Oncidium 'Sharry Baby'

Cattleya skinneri

Oncidium 'Goldiana'

Chapter 4

CLIMBERS AND CREEPERS

Plants that tend to climb vertically on to supports are called climbers. Creepers are plants that tend to creep or ramble over the ground or other surfaces spreading horizontally. These have showy flowers or decorative foliage. Climbers are conventionally used in gardens to cover fences, bare walls and tree trunks. They are used to provide ornamentation in a vertical aspect. Creepers on the other hand are used as ground covers and function as decorative elements in a horizontal aspect.

CLASSIFICATION OF CLIMBERS AND CREEPERS

Depending upon their habit, vigour, duration, nature of flowering, adaptations for climbing and shade tolerance, they can be classified as follows:

1. Based on plant habit and vigour

a) Succulent climbers: *Monstera, Philodendron, Syngonium.*

b) Light and thin-stemmed climbers: *Clitoria ternatea, Senecio confusus, Solanum jasminoides.*

c) Heavy, woody climbers: *Artabotrys, Bougainvillea, Petrea.*

2. Based on longevity and nature of flowering

a) Annual climbers: *Clitoria*, Black-eyed Susan, *Quamoclit pennata.*

b) Perennial ever-flowering climbers: *Allamanda, Antigonon, Tristellatia.*

c) Perennial seasonal flowering climbers: Cats-paw climber, Garlic creeper, *Congea tomentosa.*

d) Perennial foliage climbers: *Asparagus, Monstera, Vernonia.*

3. Based on adaptations for climbing

a) Tendril climbers: *Antigonon*, Glory lily, *Passiflora.*

b) Thorny climbers: *Asparagus, Bougainvillea*, Roses.

c) Adventitious root climbers: *Ficus repens, Philodendron, Syngonium.*

d) Twiners: *Jacquemontia*, Jasmines, *Ipomoea.*

4. Based on shade tolerance

a) Shade loving foliage climbers: *Philodendron, Scindapsus, Monstera.*

b) Shade tolerant flowering climbers: Clock vine, Potato creeper, Jasmines.

5. Based on the nature of floral fragrance

a) Strongly scented climbers: *Artabotrys*, Jasmines, *Wisteria.*

b) Delicately scented climbers: *Hiptage, Clematis, Quisqualis.*

FLOWERING CLIMBERS AND CREEPERS

Flowering climbers and creepers popularly grown in Indian gardens are listed below:

Adenocalymna alliaceum Bignoniaceae

Garlic creeper

Plants are glossy leaved evergreen climbers that prefer open sunny situations. The ovate-lanceolate leaves are trifoliate. The leaves and flowers when crushed emit a garlic-like smell. Flowering is seasonal. In peninsular

India, flowering starts from August-September and continues up to December-January. Flowers are funnel-like and are produced in dense clusters. They are mauve-pink and fade to pale pink. *Adenocalymna calycina* (syn. *Adenocalymna comosum*) has golden yellow flowers and is called the "Yellow Garlic creeper".

Allamanda cathartica Apocynaceae

Golden trumpet, Canario, Koalambipoo (Mal.), Yellow Allamanda .

Plants are ever blooming shrubby climbers of the tropics. The ovate and glossy leaves are arranged in whorls of 3 to 6. The golden yellow flowers are trumpet-like and are produced in clusters. they are single or double. *Allamanda blanchetii* 'Cherries-Jubilee' is a robust cultivar with pink to purplish-red flowers and light green downy leaves. *Allamanda schotii* called the 'Dwarf Allamanda' or 'Oleander leaf Allamanda' is short in stature and has a shrub-like habit. The leaves, branches and flowers are smaller than 'Yellow Allamanda'. The free lobes of the petals are broadly pointed in this. A cultivar named 'Australian Silver' has gray-green leaves.

Antigonon leptopus Polygonaceae

Chain-of-love, Coralita, Coral-vine, Mexican creeper.

Plants are tuberous rooted evergreen tendril climbers that prefer open sunny situations. The leaves are heart-shaped, and have attractively reticulate lamina. Flowers have showy cupped tepals that are white or light to dark pink. These are produced in paniculate racemes. *Antigonon guatemalense* called the Guatemalan Coral-vine or Coraliyto is a semi-deciduous climber. The tepals have spreading lobes and ruffled margins. The leaves are cordate and pubescent.

Arachis pintoi Fabaceae

Perennial peanut, Pinto Peanut

Plants are slender stemmed creeping perennials that are used as ground covers in tropical gardens. They tend to spread like a mat and carry yellow flowers. The leaves and flowers closely resemble groundnut but are smaller in size. The cultivar 'Golden Glory' with more profuse flowering is used ornamentally and the cultivar 'Amarillo Peanut' with less profuse flowering is used as a forage ground cover in plantations and forestry.

Aristlochia elegans Aristolochiaceae

Birthwort, Calico Flower, Hamsa kodi (Mal.), Pakhilata (Beng.).

Plants are slender stemmed evergreen climbers. The leaves are cordate. The solitary flowers have a showy calyx, which is curved and formed into a hood at one end. It is pale greenish white and has maroon-red reticulations. The flowers are not obnoxiously scented. *Aristlochia grandiflora* called the 'Pelican flower' has large swan-like flowers that emit an offensive odour.

Artabotrys (See under Flowering Shrubs for the Tropics).

Asparagus (See under Monocotyledonous Foliage Herbs)

Banisteria laevifolia Malpighiaceae

Banisteria

Plants are winter flowering woody climbers of the sub-tropics, producing golden yellow flowers in compact clusters. In *Banisteria laurifolia*, the yellow flowers are held in a spray.

Bauhinia diphylla Fabaceae

Climbing Bauhinia

Plants are robust climbers adapted to moist shady areas of tropics. The large leaves are bilobed and the flowers are yellowish white and are in clusters of 4 to 6. *Bauhinia fassoglensis* is a tuberous rooted climber having partially bilobed leaves and yellow flowers. *Bauhinia vahili* called the 'Malu creeper' is native to the foothills of the Himalayas. It is a semi-deciduous woody climber that is at times trained to grow as a shrub in Indian gardens. The flowers are white and the petal margins are attractively ruffled. *Bauhinia yunnanensis* is the 'Yunnan Bauhinia' or 'Chinese Bauhinia'. It is an evergreen tendril climber adapted to partially shaded lighted situations. The flowers are white and have magenta pink-veined petals.

Beaumontia caryophyllata Apocynaceae

Malatilata (Mal.) Madoblata (Hin.)

Plants are vigourous hardy climbers. The leaves have red-tinged veins. The flowers are white, star-like with twisted petals. The floral scent is clove-like.

Beaumontia grandiflora Apocynaceae

Nepal Trumpet climber

Plants are heavy woody climbers having shiny leaves. The flowers are trumpet-like or lily-like and faintly scented. *Beaumontia jerdoniana* is a short statured and small-leaved species having scented white flowers.

Bignonia unguis-catii Bignoniaceae

Cat's-paw climber

Plants are vigorous woody climbers native to South America. In Indian gardens they were trained over arches and pergolas during the post-colonial period. A few are found as remnants in public and institutional gardens. Golden yellow flowers cover the vines in December –January in peninsular India. Their splendour is short lived, as the flowers are ephemeral. However, the canopy of softly falling flowers and the golden carpet they spread below is a sight of unparalleled beauty.

Bougainvillea Spachs Nyctaginaceae

Bougainvillea, Paper flower

This is a genus of thorny evergreen climbing shrubs native to S. America having spectacularly coloured floral bracts. They were introduced into India in 1860 from Europe. A large number of cultivars are grown all over the world. They are hybrids or bud sports of *Bougainvillea glabra, Bougainvillea peruviana* and *Bougainvillea spectabilis*. The leaves are pubescent or glabrous. Axillary buds are modified into thorns. The floral bracts are papery in texture, broadly ovate and variously coloured. They may be bicoloured with bracts of two different colours as in 'Mary Palmer', which has magenta-pink and white bracted flowers. Bracts are three in number and enclose small tubular flowers in single cultivars. They are 20 to 40 in the multi-bracted 'Mahara' cultivars, which are sterile and do not possess true flowers. The flowers have a terminal star-like edge which maybe prominent or rarely, absent. Variegated cultivars like 'Vellayani' having creamy white and green leaves are also grown. *Bougainvillea* 'Thimma' is a variegated cultivar having green and yellow leaves and bicoloured bracts like 'Mary Palmer'. Bougainvilleas are usually trained as shrubs or standards or are pot grown in gardens. *Bougainvillea arborea* is a spine-less semi-deciduous shrub having shiny pubescent leaves. The bracts are lilac-purple in colour and are persistent.

Campsis grandiflora Bignoniaceae
Syn. *Campsis chinensis*

Chinese trumpet creeper, Orange trumpet vine

Plants are deciduous woody climbers native to China and Japan. The leaves are broadly ovate, and toothed. Aerial roots develop from the leaf axils. Flowers are reddish orange, trumpet-like and in clusters of 6 to 8.

Campsis radicans Bignoniaceae

Trumpet vine, Trumpet honey suckle

Plants are deciduous woody climbers native to South-East Asia. The leaves have axillary aerial roots. The flowers are orange, trumpet-like and are produced in elegant panicles.

Choneomorpha fragrans Apocynaceae
Syn. *Choneomorpha macrophylla*

Frangipani vine

Plants are evergreen climbers with long, broadly ovate pubescent leaves. The flowers have a short tubular throat that is yellow in colour and terminating in free lobes that are twisted and white in colour. The flowers are fragrant. The vine fibers are used as netting in Indo-Malaysian regions.

Clematis paniculata Ranunculaceae
Syn.Clematis vittalba

Clematis, Laghuparnika (Sans.), Wedding Bouquet

Plants are spreading climbers that flower profusely and intermittently in peninsular India. The flowers are white, delicately scented and are produced in panicles. *Clematis alpinii* is a spring-flowering temperate species having violet-blue flowers. *Clematis jackmanii* has large blue to mauve-pink flowers.

Clerodendrum splendens Verbenaceae
Syn. *Clerodendron superbum*

Glory Bower-vine

Plants are evergreen climbers suited for open sunny situations as well as partially lighted areas of the tropics and sub-tropics. The ovate leaves have roughly ruffled surfaces. The flowers are produced in dense clusters. They are red and have a pale green calyx. *Clerodendrum speciosum* has a maroon-purple calyx and bright red flowers. It is considered to be a natural hybrid of *Clerodendrum splendens* and *Clerodendrum thomsoniae*.

Clerodendrum thomsoniae Balf. Verbenaceae

Bleeding heart, Bag flower

Plants are evergreens having a twining habit adapted to partially shaded situations. They are also grown as sub-shrubs in the open. They have attractively crinkled dark green leaves. The flowers are blood red in colour and are held in a showy white-cupped calyx. *Clerodendrum paniculatum* Linn. having a red-tinged green calyx is more robust than *Clerodendrum thomsoniae*.

Clitoria ternatea Fabaceae

Butterfly creeper, Shankhu Pushpam

Plants are annual climbers that have small ovate leaves. The flowers are white or light or dark blue. An inky blue cultivar with double flowers is also grown.

Clytostoma callistegioides Bignoniaceae
Syn. *Bignonia callistegioides*

Argentine Trumpet vine, Love-charm, Painted Trumpet vine

Plants are evergreen climbers adapted to partially shaded situations. Leaves are trifoliate and the terminal leaflet is usually modified into a tendril. The flowers are funnel shaped, produced in twos and are mauve-pink in colour. The tubular throat is coloured creamy white and has maroon veins.

Combretum coccineum Combretaceae

Scarlet Comb

Plants are prickly evergreen climbing shrubs adapted to open and partially shaded situations. The leaves are oblong and elongated. The small red flowers have long exerted stamens. They are produced in panicles. *Combretum grandiflorum* is non-spiny and has ovate leaves with purple coloured petioles and mid-vein. The perianth segments are coloured red and the stamens are yellow. *Combretum rotundifolium* called the 'Monkey Bush' is a semi-deciduous climber producing small red flowers with yellowish orange stamens on elongated brush-like inflorescences.

Congea tomentosa Lamiaceae

Shower-of-Orchids

Plants are vigorous evergreen woody climbers that can be trained to grow as shrubs. The young shoots are downy and the leaves are ovate-elliptic. The floral bracts are showy and are coloured mauve-pink while the flowers are white and inconspicuous. Flowers are produced in December-January.

Crypotostegia grandiflora Asclepiafaceae

Rubber vine

Plants are evergreen twining climbers. The leaves are dark green, elliptic glossy and thick with light green coloured veins. The flowers are violet in colour and trumpet shaped. *Crypotostegia madagascarensis* called the 'Madagascar rubber-vine' has reddish violet flowers.

Delima sarmentosa Dilleniaceae

Delima

Plants are partially shade loving woody climbers. Terminal panicles bear white magnolia-like scented flowers during summer.

Derris scandens Fabaceae

Derris vine

Plants are vigorous woody climbers. The flowers are pale pink and are produced in profusion from March onwards.

Dissotis rotundifolia Melastomaceae

Trailing Glory Flower

Plants are succulent stemmed evergreen creepers. The leaves have red petioles. The flowers are solitary and have five violet-pink petals.

Episcia cupreata Gesneriaceae

Episcia, Flame Violet

Plants are shade adapted succulent perennial creeping herbs. The leaves are ovate, green or brown tinged, thick and coarse in texture with silvery markings on the lamina. The green-leaved plants have red flowers and the bronze-leaved plants have pink flowers. They are attractive ground covers for moist and shady areas.

Gloriosa superba Liliaceae

Glory lily, Agnishikha (Sans.), Maenthonni (Mal.)

Plants are tuberous rooted climbers. They have strong shoots with distichously held leaves and readily clamber up trees. The leaf tips are modified into tendrils. Flowering starts in August-September. The opened flowers have frilled yellow petals with scarlet tips. With ageing, the red colour extends downwards, turning the petals entirely red with a yellow central portion and prominent orange-yellow anthers.

Gmelina philippensis Lamiaceae
Syn. *Gmelina hystrix*

Hedgehog flower, Wild sage

Plants are evergreen climbing shrubs. The elliptic leaves have axillary spines. The inflorescence is pendant and the flowers are pale yellow, cup-shaped and arise in the axils of large persistent bracts that turn light brown. The flowers open in pairs.

Heterocentron elegans Melastomaceae
Syn. *Schizocentron elegans*, *Heeria elegans*, *Monochaetum guatemalensis*

Spanish Shawl

Plants are evergreen succulent creepers having reddish stems. The leaves are ovate, lightly pubescent and have serrulate margins. The flowers are produced singly. They have four magenta-pink petals with yellow stamens.

Hiptage benghalensis Malpighiaceae
Syn. *Hiptage madablota*

Madhavi-lata, Madan masta

Plants are evergreen woody climbers that are partially shade loving. The leaves are ovate-lanceolate. The flowers are fragrant, with greenish white waxy petals that are reflexed and clawed along with a smaller golden yellow petal. The flowers turn entirely yellow on ageing. They are produced in profusion from December-January onwards.

Hoya multiflora Apocynaceae
Syn. *Centrostemma multiflorum*

Shooting star flower, Wax flower

Plants are evergreen semi-epiphytic climbers adapted to partially lighted situations. The leaves are ovate, glossy and dark green. The flowers are white with reflexed perianth lobes. Flower clusters have the appearance of radiating shooting stars. *Hoya pottsii* syn. *Hoya angustifolia* has very small cream coloured flowers with a purple center. *Hoya purpurofusca* has long pendant clusters of white and pink flowers. All the above species are indigenous to Malaysia, Indonesia and South China.

Ipomoea L. Convolvulaceae

Morning Glory

The genus has a number of climbing species that are grown ornamentally throughout the world. The leaves are deeply or broadly lobed, and shiny dark green to dull gray-green. Plants are ever flowering in the tropics. The flowers are ephemeral but being funnel shaped and large in size, they lend beauty to arches, walls and pillars covered by them.

The perennial species include *Ipomoea acuminata*, the Moonflower or Good Night flower having greenish white fragrant flowers, *Ipomoea cairica* syn. *Ipomoea palmata*, the vigorously growing 'Railway Creeper' having pale to dark purple flowers, *Ipomoea carnea*, the woody climber having dark pink flowers, *Ipomoea horsfalliae* having digitate leaves and rose-pink to crimson flowers, *Ipomoea learii* having lilac or dark purple flowers and *Ipomoea paniculata* having large violet flowers.

The annual species include *Ipomoea purpurea* the common Morning Glory of sub-tropical and mild temperate regions having white to reddish purple flowers, *Ipomoea quamoclit* (syn. *Quamoclit pennata)* called the 'Cypress vine' or 'Cardinal Creeper' or 'Indian Pink' or 'Akaasha Mulla' having thin delicate stems with small finely divided pinnate leaves and small dark red flowers and *Merremia tuberosa* syn. *Ipomoea tuberosa* the 'Wood Rose' having yellow flowers and persistent dry fruits that resemble rose flowers. *Ipomoea versicolor* syn. *Mina lobata* called the 'Mina Creeper' is an annual adapted to sub-tropical and mild temperate situations. The flowers are yellow at opening and turn orange and then red.

Jacquemontia violacea Convolvulaceae
Syn. *Jacquemontia pentantha*

Indian Bluebells

Plants have thin, wiry, twining stems that turn woody. The leaves are small and broadly ovate. The flowers are small, funnel shaped and the corolla is mauve-blue.

Jasminum L. Oleaceae

Jasmines

The genus has a number of fragrant or non-fragrant shrubby woody climbers that are grown in gardens of the tropics and sub-tropics. They are traditional favourites of gardens of the East. Plants are partially shade tolerant and readily climb over supports including trees. They can also be trained to grow as shrubs in open sunny situations. Most of the species have a seasonal flowering habit. Flowering starts from December onwards in India and extends up to June. The important species grown as climbers in Indian gardens include *Jasminum auriculatum, Jasminum grandiflorum, Jasminum laurifolium, Jasminum multiflorum, Jasminum officinale, Jasminum primulinum* and *Jasminum sambac.* Plants raised from plageotropic shoots of *Jasminum sambac* have a bushy habit. Selected cultivars of this species are extensively grown in South India for commercial flower production.

Lantana sellowiana (See under Popular Flowering Shrubs of Gardens).

Lonicera hildebrandeana Caprifoliaceae

Giant Honey-suckle

Plants are woody climbers adapted to temperate regions. The fragrant flowers are creamy white and deepen to a peach colour on ageing. The

'Japanese Honey-suckle' *Lonicera japonica* is suited to partially shaded situations and has tubular white flowers with a violet tinge. The trumpet honey suckle *Lonicera sempervirens* has tubular non- scented flowers that are red on the outside and coloured golden yellow inside.

Mandevilla boliviensis Apocynaceae

White Mandevilla

Plants are evergreen climbers adapted to open as well as partially lighted situations. The leaves are elliptic and glabrous. The white flowers are bell-shaped and have a yellow throat. *Mandevilla sanderi* is called the 'Pink Mandeville' and has rough textured and slightly pubescent leaves. The flowers are funnel shaped and pale to dark pink in colour.

Nepenthes khasiana Nepenthaceae

Pitcher plant

Plants are herbaceous carnivorous climbers adapted to moist shady regions of North Eastern India. The lamina ends in a pitcher like floral pouch that is greenish and maroon in colour and has an overhanging lid. Insects slip down the slimy edge and walls and are trapped in the digestive fluids held in the pitcher. *Nepenthes rajah* is the 'Rajah Pitcher plant' that collects up to 2 liters of fluid in its pitcher.

Odontadenia macrantha Apocynaceae
syn. *Odontadenia grandiflora*, *Odontadenia speciosa*

Odontadenia

Plants are evergreen woody climbers. The leaves are dark green, ovate and leathery. The fragrant flowers produced in small clusters are funnel shaped, yellow in colour and have streaks of light orange colour.

Pandorea jasminoides Bignoniaceae
Syn. *Bignonia jasminoides, Tecoma jasminoides*

Bower plant

Plants are evergreen shrubby climbers. The leaflets are ovate and bright green in colour. The flowers are white and are coloured light to dark pink down the throat. They are found in small clusters. *Pandorea jasminoides* var. *alba* is a large flowered variety.

Passiflora coerulea Passifloraceae

Blue Passionflower

Plants are tendril climbers having glossy 5 to 9 lobed leaves. The flowers have white petals and sepals and a pale pink corona and blue stamens. *Passiflora coccinea* has entire leaves. The flowers have red petals and sepals and a red corona. *Passiflora edulis* the 'Yellow Passionflower' has yellow and purple-fruited forms. The flowers are white and the corona is coloured purple.

Passiflora foetida Passifloraceae
Syn. *Passiflora gossipiifolia*

Goat Passionflower

This climber has entire or three lobed leaves. The flowers have white tepals and a purple and white corona.

Pergularia odoratissima Asclepiadaceae

West coast creeper, Labanga latha

Plants are evergreen climbers. The flowers are fragrant, greenish yellow and are produced throughout the year.

Petrea volubilis Verbenaceae

Purple Wreath

Plants are hardy woody climbers that have ovate leaves with a rough papery surface texture. The flowers are borne in wreath-like clusters. They are violet in colour and are held within a lilac coloured showy calyx that is persistent. Plants are garden favourites of the tropics and subtropics and are trained to form arches, pergolas and standards.

Podranea ricasoliana Bignoniaceae
Syn. *Pandorea* rica / *Tecoma rica*

Pink trumpet vine, Bubble gum vine

Plants are woody climbers with light green delicate-looking leaves having serrate margins. The flowers are pink, trumpet shaped and are produced in few-flowered clusters.

Poranopsis paniculata Convolvulaceae
Syn. *Porana paniculata*

Bridal bouquet, Safed bel, Snow-in-the-Jungle

Plants are evergreen woody climbers that can be trained as shrubs. The leaves are dark green and cordate. The flowers are produced in panicles and are small, white and funnel-shaped. *Poranopsis racemosa* called the 'Snow creeper' has smaller leaves and is annual in habit. The flowers are small, white and star-shaped. *Poranopsis volubilis* is a shrubby climber that has rounded terete stems and white flowers.

Pyrostegia venusta Bignoniaceae
Syn. *Bignonia venusta*, *Pyrostegia ignea*

Flame vine

Plants are evergreen woody climbers. The leaves are trifoliate with ovate, slightly pubescent leaflets. The terminal leaflet is often modified into a tendril. The flowers are tubular, bright orange-yellow and they cover the vines almost entirely during the flowering season. In southern India, flowering is observed from September onwards in the hill zones.

Quisqualis indica Combretaceae
Syn. *Quisqualis glabra*, *Quisqualis pubescens*

Rangoon creeper, Rangoon-ki-bel, Rangoon Malli

Plants are hardy, invasive ever-flowering climbers. The flowers are fragrant and are produced in clusters. The flowers are white at opening and become pink on ageing and are nearly red when they start to fade. Entirely white or red flowered varieties are also grown.

Rosa spp. Rosaceae

Climbing Roses, Rambler Roses

Climbers and ramblers belonging to *Rosa multiflora, Rosa filipes* and *Rosa wichuraiana* are all-time favourites of temperate gardens. In the tropics and sub-tropics, these are grown as weeping standards and used to cover arches and entrances in gardens. Climbers need judicious pruning and summer protection for maintaining their leaf cover throughout the year.

Saritaea magnifica Bignoniaceae
Syn. *Arabidaea magnifica*

Campanilla, Purple Bignonia, Saritaea

Plants are evergreen climbers that thrive in open as well as partially shaded situations. The leaves are trifoliate, ovate and the terminal leaflet is often modified into a tendril. Flowers are funnel shaped, magenta-purple in colour, with a white throat.

Senecio confusus Asteraceae
Syn. *Pseudogynoxis chenopodioides*

German Ivy, Mexican Flame vine, Sunshine shelly

Plants are thin-stemmed climbers adapted to partially lighted situations. The leaves are slightly pubescent and broadly serrate. The flowers are yellowish orange at opening and become dark orange-red on ageing.

Solanum jasminoides Solanaceae

Potato vine

Plants are thin-stemmed twiners adapted to partially shaded situations. The flowers are white, with a short corolla tube and are star-like. *Solanum jasminoides* var. *grandiflorum* is more vigorous and produces flowers in larger clusters. *Solanum seaforthianum* (syn. *Solanum aureum, Solanum venustum)* has lilac flowers and *Solanum wendlandii* is a sparsely spiny vine producing pale lilac flowers in larger clusters.

Stephanotis floribunda Apocynaceae
Syn. *Marsdenia floribunda*

Bridal Wreath, Clustered Wax flower, Floradora, Madagascar Jasmine

Plants are evergreen twiners. The leaves are ovate, and dark green. The flowers are fragrant and are produced in large clusters during summer.

Stictocardia tilifolia Convolvulaceae

Spotted Heart, Stictocardia, Sugar candy

Plants are succulent stemmed evergreen climbers. The leaves are cordate, glabrous, with well-marked veins. The flowers are dark reddish-pink.

Stigmaphyllon ciliatum Malpighiaceae

Golden vine

Plants are thin-stemmed creepers having pubescent young shoots. The flowers are bright yellow and have frilled petals.

Strongylodon macrobotrys Fabaceae

Jade vine

Plants are hardy woody climbers with trifoliate leaves. The flowers are greenish blue and are borne on very long inflorescences attaining a length of up to 1.5m.

Strophanthus gratus Apocynaceae
Syn. *Roupellia grata*

Climbing oleander

Plants are semi-deciduous vigorous woody climbers. The leaves are obovate, glabrous and dark green. The scented tubular flowers are white to pink in colour with broadly spiraling free lobes having wavy margins. Plants are often trained as shrubs and standards. A dark pink corona is present.

Strophanthus preussii Apocynaceae

Medusa flower

Plants are semi-deciduous shrubby climbers that can be trained as shrubs. The ovate glabrous leaves have short petioles. The tubular flowers have free corolla lobes with thread-like tips that hang twisted. *Strophanthus divaricata* has green lines along the corona and short petal tips.

Thunbergia alata Acanthaceae

Black-eyed Susan

Plants are thin-stemmed climbers adapted to partial shade. The flowers have a short corolla tube and are golden yellow with a prominent black dot at the center.

Thunbergia fragrans Acanthaceae

Clock vine, Heavenly Blue

These are attractive climbers suited to the humid tropics. The leaves are dark green with broadly serrate margins. The flowers are coloured white, pale blue or lilac and held in elegantly drooping clusters.

Thunbergia mysorensis Acanthaceae

Mysore Clock vine, Mysore-Thunbergia

These are hardy vines adapted to flower in temperate situations. The flowers are spectacularly coloured orange and maroon-red and hang in long racemes.

Torenia glabra Scorphulariaceae

Creeping Torenia

Plants are shade loving tender stemmed perennial creepers adapted to the tropics. The flowers are lilac to violet coloured.

Tristellateia australasiae Malpighiaceae
Syn. *Galphimia glauca*

Shower-of-Gold vine

Plants are evergreen and ever flowering climbers with attractive ovate shiny leaves. The flowers are small, star like and yellow, produced in rounded clusters.

Wedelia trilobata Asteraceae
Syn. *Sphagneticola trilobata*

Creeping daisy

Plants are sun loving vigorous creepers that tend to become invasive. The leaves are tri-lobed towards the upper half. The flowers are small, yellow, daisy like and solitary.

Wisteria chinensis Fabaceae

Chinese Wisteria

Plants are heavy woody climbers suited for temperate situations. They produce flowers in profusion in summer. The flowers are mauve to deep lilac.

FOLIAGE CLIMBERS AND CREEPERS

Foliage Climbers and creepers that are primarily grown for their attractive leaves and stems are listed below: -

Asparagus spp. (See under Monocotyledonous Foliage Herbs)

Callisia elegans Commelinaceae

Callisia

Plants are delicate succulent stemmed shade-loving creeping perennials having small, ovate to heart shaped, closely spaced leaves with violet colour on the lower surface. They are ideal for hanging pots.

Centella asiatica Apiaceae

Centella, Kudangal

Plants are succulent trailing herbs adapted to open moist situations. The leaves are small, thick, fleshy and sub-orbicular in shape with broadly crenulate margins. They can be used to provide a bright green ground cover in frequently wet garden areas. They root readily at the nodes and spread quickly in ideal situations. The flowers are inconspicuous and reddish purple in colour.

Cissus discolor Vitaceae

Rex-Begonia vine

Plants are tender stemmed shade loving tendril climbers. The leaves are dark green with dark violet and gray-white patterning resembling Rex-begonias.

Cissus rhombifolia Vitaceae

Grape Ivy

Plants are tendril climbers with attractive rhomboid leaves. They are used making for shady arbours and are also grown as houseplants.

Commelina benghalensis Commelinaceae

Commelina

These are shade adapted succulent stemmed creepers. The leaves are ovate, green and often faintly lined with light green or white colours.

Epipremnum (see under Monocotyledonous Foliage Herbs)

Ficus radicans Moraceae

Creeping Fig

Plants are evergreen creepers suited for hanging baskets. The leaves are dark green and often variegated.

Ficus repens Moraceae

Indian Ivy

These are evergreen creepers used to cover bare walls in gardens. The leaves are small and ovate in shape. The stems have aerial roots. Plants can also be trained as shrubs.

Gynura sarmentosa Solanaceae

Velvet plant

These are succulent shade adapted creepers. Soft violet hairs cover the whole plant body. The flowers are small and orange coloured.

Hedera helix Araliaceae

Common Ivy, English Ivy

These are vigorous climbers. The juvenile leaves and those of the climbing stems are palmately lobed while the leaves of the flowering shoots are cordate. A number of varieties with white or yellow variegations are grown.

Hemigraphis colorata Acanthaceae

Hemigraphis, Murivu-koodi

Plants are thin-stemmed creepers adapted to shady situations. The leaves are violet in colour and have a silvery surface sheen.

Ipomoea vareigata Convolvulaceae

Ornamental Sweet potato

Plants are vigorous leafy creepers adapted to the tropics. The leaves are coloured an unusually bright light green in cv. 'Margarita' and are coloured light green, pink and white in cv. 'Tricolor'. In the humid tropics they can be grown as semi-perennial evergreens for a few seasons continuously. While in the sub-tropics, the leaves dry out in winter.

Monstera (see under Monocotyledonous Foliage Herbs)

Pellionia repens Urticacae

Pellionia

Plants are shade-loving creepers and have ovate green leaves that are faintly marked with a brown coloration. They are ideal for hanging pots.

Philodendron (see under Monocotyledonous Foliage Herbs)

Piper ornatum Piperaceae

Spotted pepper

Plants are shade loving evergreen climbers native to the East Indies. The green leaves are pink spotted when young and the spots become white with age.

Sedum brevifolium Crassulaceae

Creeping Sedum

Pants are succulent-stemmed and adapted to partial shade. The leaves are small thick and fleshy. They are sickle shaped and are held on long thin branches.

Syngonium (see under Monocotyledonous Foliage Herbs)

Vernonia elaegnifolia Solanaceae

Curtain plant

These are quick growing evergreen climbers having thin shoots that carry leaves in two opposing ranks. The shoots after climbing on to supports, hang gracefully downwards. The flowers are small and greenish white.

contd. colour plates

Cryptostegia grandiflora

Bignonia unguis-catii

Cat's paw climber on a leafless Mahogany

Jacquemontia violacea

Thunbergia grandiflora

Tristellateia australasiae

Clematis vittalba

Impmoea quamoclit

Petrea volubilis

Senecio confusus

Strophanthus gratus

Thunbergia alata

Antigonon leptopus

Gloriosa superba

Ipomoea purpurea

Quisqualis indica

Stictocardia tilifolia

Torenia glabra

Callisia elegans

Centella asiatica

Cissus discolor

Episcia cupreata

Hemigraphis colorata

Pilea cadieri

Chapter 5

BULBOUS PLANTS

Among garden plants, herbaceous perennials that produce perennating structures like bulbs, corms, tubers, rhizomes or fascicled tuberous roots are called bulbous plants. They belong to various plant families such as Amaryllidaceae, Cannaceae, Heliconiaceae, Iridaceae, Liliaeceae, Musaceae, Ranunculaeceae and Zingiberaceae. The possession of perennating structures that are modifications of plant parts- mostly stems- is common to these plants. These modifications aid in their vegetative propagation as the dormant buds associated with them regenerate when conditions are favourable to result in new plants. In both tropical and temperate climate zones, bulbous plants have evolved and naturalized in several ecological 'niches'. Many of these have been domesticated centuries ago and are familiar garden plants. Generally, bulbous plats are adapted to grow vegetatively putting forth shoots carrying a limited number of leaves. When further vegetative growth has ceased these shoots produce terminal or basal inflorescences. Bulbous plants are grown and displayed in clumps.

Based on the requirement of relatively low or high temperatures for sprouting and flowering they are classified into **warm- season bulbous plants** and **cool-season bulbous plants**. These groupings are not exclusive and many plants have evolved varieties that thrive in both seasonal conditions.

WARM SEASON

The following are the **warm-season bulbous plants** that are grown in tropical gardens:

Alocasia (Schott.) G. Don — Araceae

Alocasia

Plants are evergreen herbs having attractive sagittate leaves. The leaves are dark green with a metallic surface sheen and grayish-white coloured well marked veins. The lamina has scalloped and pointed basal lobes in the cultivar 'Amazonica'. In these, the lamina attains a length of up to 5m and a width of up to 3m. Plants are considered to be natural hybrids of *Alocasia sanderiana* Hort. ex Bull and *Alocasia longiloba* Miq.

Alocasia cuprea K. Koch — Araceae

Giant Caladium, Alocasia

Plants have ovate leaves with partially united basal lobes. The leaf surface is closely veined with a darker green colour and has a corrugated appearance. The lamina has an overall metallic-green colour on the upper surface and the lower surface is purple-tinged.

Alocasia longiloba Miq. — Araceae

Syn. *Alocasia lowii* Hook.

Alocasia

Plants are evergreen herbs with thickened corm-like stems. The leaves are elongated, sagittate and the basal lobes as well as the top are pointed. The main veins as well as their feathery branches are marked silvery white on the lamina.

Alocasia macrorhiza Schott. cv. Variegata — Araceae

Syn. *Alocasia macrorrhizos* (L.) G. Don, *Colocasia indica* Hassk.

Variegated Giant Taro, Variegated Giant Alocasia, Elephant ear.

Plants are robust herbs having elongated corm-like underground stems that gradually become visible above ground as plants age. The leaves are large, up to 0.5m in length and have patches of creamy white colour on the lamina. The corms are edible. Cultivated varieties are differentiated by the differences in their leaf colouration.

Alpinia purpurata K. Schum. Zingiberaceae

Pink ginger, Red ginger

Plants are evergreen, tall growing rhizomatous herbs adapted to partially shaded and humid situations. The linear ovate leaves are held distichously. The lamina attains a length of 30cm or more and a width of up to 15cm. Clumps with several shoots carrying brightly coloured inflorescences is particularly attractive. The shoots arising from rhizomes are thick and attain a height of more than 2m under shade. The inflorescence is fastigate, solitary or rarely branched and arises terminally. Plantlets sprout directly on the inflorescence. They root readily when the inflorescence arches down to the ground due to their weight. The flowers are small, white and are produced in the axils of pale pink or pinkish-red bracts.

Alpinia zerumbet (Pers.) B. L. Burtt & R. M. SM Zingiberaceae
syn. *Alpinia nutans* Rosc., *Alpinia speciosa* K. Scuhm.

Shell ginger

Plants are robust evergreen rhizomatous herbs adapted to moist partially shaded situations. The leaves resemble those of *Alpinia purpurata* and are broadly ovate. The inflorescences are pendant spikes produced terminally on the shoots. The flower buds are globose with pointed tips and are pale pink in colour with darker pink tips. The open flowers have a showy spoon-shaped lip that is coloured golden yellow with dark red veins. *Alpinia zerumbet* 'Variegata' is a cultivar that is shorter in stature and has leaves with bold yellow and green lines on the lamina and sheaths. The plants seldom flower when grown in containers under shade. They are grown as foliage plants.

Amaryllis belladonna L. Amaryllidaceae

Belladonna lily

Plants are native to South Africa and have spread throughout the gardens of the tropics and subtropics. They are bulbous herbs with leaves that are strap- like and up to 40cm or more in length. The tops dry out in summer before the scapes emerge. The scapes are solid, stout and 50 to 60cm or more in length. Up to 12 trumpet-shaped flowers are produced per scape. The flowers are more than 10cm long and 5cm wide and are coloured dark pink to white.

Caladium hortulanum Birdsey Araceae
Syn. *Caladium bicolor* Vent.

Caladium, Angel wings, Elephant ear

Plants are shade-loving herbs producing root tubers. The leaves and petioles die out in summer and regenerate with the monsoon in May-June. The leaves are membranous, thin and sagittate and spectacularly coloured dark pink at the center and green towards the margins.

Caladium lindeni Hort ex Engl. Araceae

White Caladium, White elephant ear

Plants have dark green leaves having white colouration in between the veins. The leaves are relatively thicker and the plants are hardier than the other two species. The mid ribs and main veins of the leaves are white in colour. A small leaved and dwarf-statured cultivar called 'Silver Caladium' has prominently white leaves.

Canna generalis L. H. Bailey Cannaceae

Canna lily, Kaana Vazha

Plants are rhizomatous herbs adapted to full sunlight and moist situations. They have attractive dark green boat-shaped leaves that are held on a stout culm. The lamina is slightly cupped. A thin maroon-red line is seen along the margins. The flowers are asymmetrical with 3 showy sepals and 3 smaller petals. The flower colours vary from crimson- red to yellow, orange and pink. Bicoloured flowers with streaks and stripes in contrasting colours are also seen. Plants are heavy feeders and are ideal for massing in open situations. A number of cultivars, varying in stature size and floral characteristics are grown in gardens.

Canna edulis Ker Gawl. Cannaceae
Syn. *Canna indica* Ruiz & Pav.

Arrowroot Canna, Indian shot, Yuquilla

Plants are rhizomatous herbs adapted to open and partially shaded situations. The psuedostem and leaves are green with a red tinge. The leaf margins are slightly wavy. The rhizomes are edible and have been used as traditional food by tribal communities. The flowers are small and red or yellow in colour. The fruit is a tubercle-like capsule containing hard rounded black seeds that are used as slingshots.

Costus speciosus Sm. Zingiberaceae

Crape ginger, Malay ginger, Channa koova

Plants are succulent stemmed rhizomatous herbs adapted to moist shady situations. The leaves are dark green, maroon-red on the under surface and arranged in a wide spaced spiral on the shoots. The flowers are white, produced in terminal leaf axils of shoots in small clusters. They are protected by reddish coloured bracts. The petals are ruffled and showy. The rhizomes have medicinal uses. The cultivar 'Variegata' has creamy white variegations on the leaves and reddish shoots. The leaves are smaller and linear-ovate in shape and are held on shoots that are attractively spiralled.

Costus malortieeanus H.Wendl. Zingiberaceae
Syn. *Costus elegans* Hort ex. Petersen

Stepladder plant

Plants are more ornamental than the above species and have broadly ovate light green coloured leaves. The leaves are finely hairy on the upper and lower surface and are arranged spirally on the shoots. Plants are attractive foliage plants suited for outdoor and indoor display.

Crinum asiaticum L. Amaryllidaceae
Syn *Crinum procerum* Carey ex Herb.

Crinum lily, Giant crinum, Poison bulb

Plants are robust, hardy, tall growing bulbous herbs attaining heights of up to 2m or more. The leaves are broadly lanceolate, up to 1m or more in length and 10 to 15cm in width with slightly wavy margins. The inflorescences emerge from the sides of the pseudostem. Their stalks are solid and they carry terminal clusters of faintly fragrant white flowers with thin and long corolla lobes. The corolla lobes are maroon-red on the outside and white coloured on the inside. Plants tend to form large clumps.

Crinum angustum Steud. is shorter in stature with smaller leaves that are flaccid.

Curcuma zedoaria Rosc. Zingiberaceae
Syn. *Curcuma inodora* Blatt.

Zedoary, Kasthuri Manjal (Mal.)

Plants are rhizomatous herbs in which the tops dry out during summer or the dry period of the year. The rhizomes have medicinal uses and are collected on a large scale from naturalized forest areas. Plants are grown as ornamental foliage plants and seasonal flowering plants in the tropics. They are grown in clumps beneath tree canopies. The leaves have a maroon-black colour along the mid-rib. After the tops dry out, the mother rhizomes left below ground produce inflorescences towards the end of the dry period followed by sprouts. The inflorescences are showy as the bracts are coloured dark pink-mauve. The bracts are spoon-shaped. The basal bracts that enclose the fertile flowers are less showy while the upper bracts that enclose sterile flowers are more ornamental. The inflorescences are marketed as 'Ornamental gingers' and used as cut flowers. The vernacular name of the plant is a misnomer based on this leaf colouring. However this has led to it being used extensively for adulterating the medicinally valuable *Curcuma aromatica*.

Dahlia variabilis Desf. Asteraceae

Dahlia

Plants are tuberous rooted succulent stemmed perennials. The leaves have broadly serrate margins. Flower heads are produced terminally and in the upper leaf axils of shoots. They have colourful ray florets and yellow disc florets. The rays differ in size, colour, shape and form and based on their number and nature, flowers are classified into singles, collarette, anemone-flowered, cactus flowered etc. Plants are propagated from tubers. New shoots arising from the upper stem end of the tubers can be separated when they are about 10cm in length and rooted for raising plants. In areas with mild winters the tubers can be left outdoors but in regions with frost and severe winters they are to be lifted and stored indoors until they show signs of sprouting.

Etlingera elatior (Jack) R. M. Sm. Zingiberaceae

Torch ginger

Plants are robust evergreen rhizomatous herbs attaining heights of more than 3m. They are adapted to partially shaded and humid situations. The pseudo stems that arise from the rhizomes are thick and carry distichous

Alpinia-like leaves. Inflorescences are stout scapes that emerge from the base of the clump. They are spectacular in colour and form, resembling a globose head inflorescence. The inflorescence consists of closely packed bracts that are coloured light pink, red or purple enclosing small yellow flowers. The bracts are clearly edged with a fine white line. They are marketed as cut flowers.

Gladiolus grandiflorus Andrews Iridaceae

Gladiolus, Sword lily

Plants are corm-producing herbs that are grown as annuals. They are planted in groups in the ground or in pots for mass display in gardens. The leaves are produced in a fan-like manner. The inflorescence emerges terminally and is erect and carries a number of florets on a long stalk. The flowers range in colour from pure white to crimson red and deep violet with a wide range of shades in between. After harvesting the inflorescence for cut flowers or after flowering has completed and the tops dry out, the corms are lifted and stored indoors for the next season's planting.

Hedychium coronarium J. Koenig Zingiberaceae

White Ginger lily, Butterfly ginger, Garland flower, Kalyana Sowgandhikam

Plants are rhizomatous herbs growing up to a height of 0.75m and are adapted to partially shaded situations. They are ginger like in habit with broader and lighter green leaves. The leaves are held distichously on the pseudo stem. Pure white fragrant flowers are produced terminally in small clusters on the leafy shoots from August- September onwards. Cutting down the flowered shoots encourage flowering and production of new shoots. It also helps to contain the clumps.

Hedychium flavescens Lodd. Zingiberaceae

The Yellow Ginger lily

Plants are adapted to temperate situations. They are found naturalized in N.E. India. They resemble the white ginger lily in habit. Flowering starts when temperatures fall, before the onset of the cold season.

Heliconia L. Heliconiaceae

Heliconia, Parrot flower

Plants are perennial evergreen rhizomatous herbs attaining heights of upto 2 to 4m or more and are adapted to moist situations having varying light levels. They produce spectacular inflorescences having bracts coloured bright yellow, orange-red, crimson, light pink and combinations of these colours with green. They are naturalized in tropical America and the South Pacific island of New Guinea. Species variability as well as varietal wealth abound in these regions. The leaves are distichously arranged or cannaceous or musoid. The inflorescences are erect and distichous in *Heliconia bihai* J. Mill. (Erect Lobster Claws), *Heliconia caribaea* Lam. (Wild Plantain), *Heliconia champneiana* Griggs, *Heliconia lingulata* Ruiz & Pav., in *Heliconia psittachorum* and in *Heliconia stricta* Huber. They are erect and spirally twisted in *Heliconia augusta* Vell., *Heliconia aurantiaca* Gsiehb., *Heliconia latispatha* Benth., *Heliconia sarapiquensis* G. S. Daniels & F. G. Stiles and *Heliconia spissa* Griggs. They are pendant in *Heliconia chartacea* Lane ex Souza Barreiros, *Heliconia collinsiana* Griggs, *Heliconia griggsiana* L. B. Sm., *Heliconia mutisiana* Cuatrec., *Heliconia pogonantha* Cufod., *Heliconia rostrata* Ruiz & Pav., *Heliconia vellerigera* Poepp. and in *Heliconia xanthovillosa* W. J. Kress. *Heliconia wagneriana* Petersen called the 'Scarlet Banana has large beautifully sculpted bracts that are coloured with a mixture of bright green, yellow and orange-red. *Heliconia indica* Lam. with coppery green musoid leaves and *Heliconia metallica* with cannaceous leaves that are maroon-purple on the under surface are the two species grown as ornamental foliage plants.

Hippeastrum reticulatum Herb. Amaryllidaceae

Knight's star, Azucena, Hippeastrum lily

Plants are evergreen bulbous herbs having short strap-like leaves that taper broadly. Leaves have a white line along the midrib in *Hippeastrum reticulatum* var. *striatifolium* (Herb.) Herb. The flowers are pale pink with fine red lines all over the corolla or orange-red creamy white in colour.

Hippeastrum reginae Herb.is called the 'Mexican lily' and has bright red flowers and a pale white star down the throat of the corolla tube. Plants are ideal for pot display, for clump planting and also for edging walks and lawns.

Hymenocallis littoralis Salisb. Amaryllidaceae

Spider lily

Plants are evergreen bulbous herbs adapted to partially lighted situations and attain heights of up to 0.5m. The flowers are white in colour. The leaves are narrow, strap-like and arch gracefully. The corolla is united at the base and the circular margins have intermittent thread-like extensions that arch outward resembling spider legs.

Kaempferia pulchra Ridl. Zingiberaceae

Peacock Ginger lily

Plants are low-growing evergreen rhizomatous herbs that thrive in shaded and humid situations having slightly alkaline soils. The leaves are broadly ovate with dark green patches on the lamina. The flowers are pink-mauve with slightly ruffled petals. Plants are ideal ground covers for moist shady locations.

Musa ornata Roxb. Musaceae

Flowering banana, Ornamental banana

Plants are short statured perennial herbs with rhizomatous stems. The pseudo stems are thin with few leaves. In open situations plants form small clumps. The male buds are erect and the bracts are attractively coloured pink, light orange, light purple or white streaked. The fruits are also ornamental, being brightly coloured, small, and seedy.

Neomarica caerulea (Ker Gawl.) Sprague Iridaceae
Syn.*Cypella caerulea* Seub. ex Hook. f.

Apostle plant, Marcia iris, Lirio

Plants are rhizomatous evergreens attaining a height of up to 0.5m. They have equitant (fan-like) sessile leaves with sharply pointed tips arising directly from the rhizome. Inflorescences arise from the mid rib of leaves. The flowers are white with bluish markings. Plants are clump forming and are displayed in pots.

Polianthes tuberosa L. Amaryllidaceae

Rajanigandha (Beng., Hindi), Gul-e-shabu (Urdu), Sugandharajan (Kan., Tam.)

Plants are low growing bulbous herbs attaining a height of 15 to 20-cm. The leaves are grass-like and are produced in a basal whorl. They arch gracefully, attaining a length of 30cm or more. Inflorescences arise terminally and are nearly 1m in length. The flowers are white, single, semi double or double and very fragrant. They have a tubular corolla with free upper lobes.

Proiphys amboinensis Herb. Amaryllidaceae
Syn. *Eurycles amboinensis* Lindl., *Eurycles sylvestris* Salisb.

Cardwell lily, Eurycles lily

Plants are bulbous herbs that are short in stature attaining a maximum height of 0.5m. They are adapted to shaded and partially shaded situations. The flowers are small, white and in groups of ten or more. The leaves are rounded to reniform. The deeply lined veins give the lamina a corrugated appearance. The fruits are rounded and have viviparous seeds.

Scadoxus multiflorus Raf. Amaryllidaceae
Syn. *Haemanthus multiflorus* Martyn., *Haemanthus katherinae* Baker, *Haemanthus kalbreyeri* Baker, *Haemanthus tenuiflorus* Herb.

Blood lily, Blood flower, Football lily, Red Cape lily

Plants are bulbous herbs attaining a height of up to 0.5m. They thrive in partially lighted situations. The leaves are strap-like. The tops die out in summer and the scapes alone emerge carrying a globose head of radiating bright red flowers.

Zephyranthes L. Amaryllidaceae

Fairy lily, Zephyr lily

These are the smallest among the bulbous plants grown in gardens. They thrive in full sunlight and partially lighted areas. The leaves are narrow, grass-like and shiny. The flowers are funnel shaped with corolla lobes that are united at the base and free at the top. Plants are designated as different species based on flower colour. *Zephyranthes ajax* Sprenger has yellow flowers, *Zephyranthes candida* Herb. has white flowers and in *Zephyranthes rosea* Lindl. the flowers are pink. Plants having larger sized pink flowers with longer leaves are grouped under *Zephyranthes grandiflora* Lindl. (syn. *Zephyranthes carinata* Herb. / *Zephyranthes tsouii* Hu).

COOL-SEASON

The following are the **Cool-season Bulbous Plants** that are popularly grown in Indian gardens.

Achimenes longiflora Gesneriaceae

Achimenes, Magic flower, Nut orchids, Widow's tears

Plants are tuberous rooted and have petunia-like flowers that are violet-blue in colour. *Achimenes grandiflora* has reddish purple flowers.

Acidanthera bicolor Iridaceae

Scented gladiolus, Peacock orchid

Plants have corm-like under ground stems. The flowers are scented, white in colour and have a maroon-red star-shaped colouration at the throat.

Agapanthus umbellatus Liliaceae

Blue African lily, Lily of the Nile

Plants have long graceful leaves of up to 50cm length. The flowers are generally blue in colour, funnel shaped and are produced in large clusters.

Allium Liliaceae

Flowering Onion, Allium

A number of flowering onions having blue, white, yellow or pink coloured flowers are grown for flowers in summer. The inflorescences are umbels with many flowers.

Amaryllis (see under Warm-season Bulbous Plants)

Anemone coronaria Ranunculaceae

Poppy Anemone, Windflower

Plants are tuberous rooted ad have finely divided leaves. The flowers have a wind-blown appearance and are produced in summer. They are poppy-like and coloured white, pink, red or purple. *Anemone fulgens* and *Anemone hortensis* have less finely divided leaves and red or mauve-purple flowers.

Arisaema speciosa *Araceae*

Snake lily, Cobra lily, Green dragon

Plants have a single leaf and a brown and green spotted petiole. The spathe is boldly coloured with green and dark brown stripes.

Belamcanda chinensis Iridaceae

Blackberry lily, Leopard flower

Plants are rhizomatous in habit attaining a height of up to 90cm. They have iris-like leaves. The flowers are 4 to 5 cm in diameter, orange with crimson spots. Plants are propagated by division of clumps.

Begonia bicolor Begoniaceae

Tuberous rooted begonia

Plants are short in stature, shade loving and adapted to low temperatures. The flowers are rose-like, delicate in texture with plain or fringed edges. They are white, pink, red, orange or yellow. The stems are thick, succulent and decumbent resembling rhizomes and are surface rooting. After flowering in August- September, the leaves dry-out and the tuberous stems are lifted and stored indoors like bulbs or rhizomes for planting in Feb- March.

Caladium (see under Warm-season Bulbous Plants)
Canna (see under Warm-season Bulbous Plants)
Clivia miniata *Amaryllidaceae*
Syn.Imantophyllum miniatum

Kaffir lily

Plants are bulbous evergreens. The flowers are funnel shaped and orange-yellow or red. They are borne in umbellate clusters. Plants are also grown indoors.

Convallaria majalis Liliaceae

Lily of the Valley

Plants are rhizomatous evergreens. The flowers are fragrant, bell-shaped and white or pink.

Crinum (see under Warm-season Bulbous Plants)

Crocus sativus Iiridaceae

Saffron crocus, Crocus, Kesar

Ornamental crocuses are grown in rockeries and are mass planted in lawns. They are autumn flowering or spring flowering. Plants are propagated by corms and need lifting and replanting once in 4 to 5 years only. Flowers are larger than those of commercial saffron and are white to dark purple.

Cyclamen persicum Primulaceae

Cyclamen, Saw board

Plants are herbs producing corms or tubers and are shade-loving. The flowers are butterfly-like with entire or frilled petals.

Dahlia (see under Warm-season Bulbous Plants)

Eucharis grandiflora Amaryllidaceae
syn. *Eucharis amazonica*

Star of Bethlehem, Amazon lily

Plants are low growing bulbous herbs with ovate-lanceolate leaves. The flowers are borne on long scapes. They are white, fragrant and 6 to 7 in number on an umbellate cluster. Plants are suited for semi-shaded and humid situations.

Freesia hybrida Iridaceae

Freesia

Plants grown in gardens are hybrids of a number of species such as *Freesia armstrongii*, *Freesia leichtlinii* and *Freesia refracta*. The flowers are funnel shaped and coloured blue, white, pink, red, orange or yellow and are produced in a raceme.

Fritillaria imperialis Liliaceae

Crown Imperial

Plants attain a height of up to 1m and have narrow leaves arranged in whorls. The flowers are yellow orange or red in colour and have a foxy odour.

Gladiolus (see under Warm-season Bulbous plants)

Gloriosa superba Liliaceae

Climbing lily / Glory lily / Maenthonni (Mal.)

Plants are tuberous rooted tendril climbers. They are grown and used for medicinal purposes in India. A number of improved garden cultivars including Dutch hybrids are popularly grown in other countries. These have flowers that are bright yellow and red to maroon red.

Hemerocallis fulva Liliaceae

Day lily

Plants are hardy perennials attaining a height of up to 90cm. The roots are tuberous. The leaves are bushy and grass-like. The flowers are single, funnel-shaped and are borne on a branched inflorescence axis. They are coloured yellow. *Hemerocallis aurantiaca* has flowers that are orange and purple. Plants are propagated by division of clumps.

Hippeastrum (see under Warm-season Bulbous plants)

Hosta plantaginea Liliaceae

Syn. *Funkia plantaginea*

Plantain lily

Plants are hardy rhizomatous perennials that thrive in shady moist situations. The leaves are tufted and flowers are white and fragrant. Garden hybrids having larger flowers in various shades are also grown.

Hyacinthus orientalis Liliaceae

Hyacinth

Hyacinth plants are bulbous perennials that are grown as annuals. Flowers are produced during spring and summer. The inflorescences are racemes carrying 30 to 40 bell-shaped florets that are white, yellow, pink, mauve, red or blue.

Iris germanica Iridaceae

Bearded Iris, Iris, Flag

Plants are dwarf to tall in habit, ranging in height from 15cm to 1m. They may be bulbous or having rhizomatous stems. Most of the English,

Spanish and Dutch irises having white, blue or yellow flowers are hybrids of a number of species such as *Iris angelica, Iris hispanica* and *Iris xiphium.* Himalayan irises like *Iris kumaonensis* and *Iris kashmiriana* have white, lilac or purple flowers and are rhizomatous.

Kniphofia uvaria Liliaceae
syn. *Kniphofia aloides, Tritoma uvaria*

Red-hot poker, Torch lily, Flame flower

Plants attain heights of up to 1.2m. They are rhizomatous with reed-like leaves. The flowers are tubular, bright red and arranged neatly and densely on the upper one-third portion of thick peduncles. As the flowers age, the colour of their lower parts lighten to orange and then to yellow. The inflorescences that are elongate to club-like in shape thus attain a spectacular colouring. Plants need not be disturbed frequently.

Lilium L. Liliaceae

Lily, Easter lily, Oriental lily

A large number species of lilies are found naturalized in the temperate zones of Asia and Europe. The European, Asiatic and oriental species have been widely hybridized to evolve varieties that combine their floral characteristics. Hence together with the white flowered varieties of Easter lily, a number of hybrids having orange, scarlet, green, yellow and crimson-purple flowers that are variously spotted and striated are grown as cut flowers and as garden-ornamentals.

Narcissus cyclamineus Amaryllidaceae

Daffodil, Narcissus, Nargis, February gold

Plants are short in stature and attain heights of up to 50cm. The leaves are narrow and grass-like. The flowers are produced singly or in clusters of up to 10 or more. The floral organs are differentiated into a central corona that is generally cupped and an outer portion made up of surrounding perianth segments. They are white, yellow, pink or red and with combinations of these colours. Daffodils are garden favorites of the temperate regions and are mass planted in lawns or pots for early spring flowers.

Ornithogalum umbellatum Lilliaceae

Star of Bethlehem

These are bulbous plants suited for partially shaded situations. The flowers are star-shaped and borne on a raceme consisting of 12 to 20 flowers. They are unusually coloured with green and white stripes on the outside and white on the inside. Plants are propagated by offsets and bulb-lets.

Oxalis regnellii 'Atropurpurea' Oxalidaceae

Purple Shamrock

Plants are dwarf in stature and have trifoliate leaves shaped like butterfly-wings. They are hardy and thrive in semi-shaded locations. The flowers are tubular, white and small in size. Plants are propagated by bulbs or seeds.

Polianthes tuberosa (see under Warm-season Bulbous plants)

Ranunculus asiaticus Ranunculaceae

Buttercup, Ranunculus

Plants are dwarf in stature with attractively cut foliage. The most common flower colour is yellow. Present day cultivars grown in gardens have white, orange, purple or red flowers. The tubers are claw-like and are lifted after flowering and stored. Plants flower from July to September.

Siningia speciosa Gesneriaceae
syn. *Gloxinia speciosa*

Gloxinia

Plants are short in stature, shade loving and have attractive leaves. The flowers are tubular and are produced in small clusters. They are white, pink, red, purpel or blue with or without an edging of white or red. Tubers or stem cuttings are used for propagation. Plants are popularly grown indoors.

Tulipa biflora Liliaceae

Tulip

Plants are short in stature and have four linear lanceolate leaves. One to four characteristically cup shaped flowers are borne centrally. They have a wide range of colours. They are important cut flowers and are also ideal for mass display in beds and containers. Planting is done towards the end of winter and flowering starts in April and continues up to June. Bulbs and bulblets are the planting materials commonly used.

Zantedeschia aethiopica Araceae
Syn. *Richardia africana*

Arum lily, Calla lily, Lily of the Nile

Plants are low growing, rhizomatous aroids having arrowhead shaped leaves. The flower is an attractive half- open spathe that is coloured white. It attains a length of up to 25cm. The tip of the spathe is narrow and pointed and slightly reflexed. Plants having yellow flowers are included under *Zantedeschia elliotiana* and those having pink flowers are grouped under *Zantedeschia rehmannii*.

contd. colour plates

Alpinia purpurata

Canna generalis

Curcuma zedoaria

Etlingera elatior

Hedychium coronarium

Caladium hortulanum

Crinum angustum

Crinum asiaticum

Hippeastrum reticulatum

Polianthus tuberosa

Proiphys amboinensis

Heliconia 'Golden Torch'

Heliconia collinsiana

Musa ornata

Musa coccinea 'Scarlet Banana'

Heliconia psittachorum 'Parakeet'

Alpinia zerumbet

Heliconia 'Pedro Ortis'

Chapter 6

SUCCULENTS AND CACTI

These are a broad group of plants that have evolved in situations characterized by very high day temperatures and low night temperatures with very low available moisture in the growing medium. Their plant body is modified to conserve moisture with physiological adaptations that help to reduce moisture loss considerably and to survive in hot desert situations or drought conditions. Stems, leaves and roots of these plants are fleshy and rich in watery mucus. The unusual morphological features that these plants have developed to enable them to survive in their harsh native habitats, set them apart from other plants and have given them rarity value when grown as ornamentals. They are hence displayed individually or grown in a rockery where a landscape of rocks and stones highlight their features and at the same time provide desirable contrast with other garden plants. Many members of Cactaceae, Crassulaceae, Aizoaceae, Euphorbiaceae, Asphodelaceae and Agavaceae belong to this group and are commonly referred to as **rock garden plants**.

SUCCULENTS

Popularly grown **succulents** of gardens are listed below:

Aeonium arboreum *Crassulaceae*

Tree Aeonium, Green Zinnia

Plants are succulent much branched shrubs that attain heights of up to 1m. The strap-like glabrous leaves are clustered in rosette-like whorls near the tips of branches and resemble zinnia flowers. *Aeonium arboreum* var. *atropurpureum* has greenish maroon leaves.

Aeonium canariense Crassulaceae

Canary Aeonium

Plants are similar in habit to the above species. The leaves are pubescent, elongated and ovate in shape with the mid-rib extended beyond the lamina like a small spur at the tip. The flowers are small and white.

Aeonium holochrysum Crassulaceae

Velvet rose

The plants resemble the above species in habit. The elongated leaves are arranged in broad rosettes. The leaves are mucronate as in the above species and the older leaves are reddish in colour. The flowers are small, yellow in colour and are held in conical erect inflorescences.

Agave L. Agavaceae

Century plants

These are attractive succulents with thick and fleshy leaves. Some species have edible plant parts and others yield fibre. *Agave americana* has broad reflexing leaves and marginal spines. *Agave americana* var. *marginata* has a marginal yellow line and var. *medio-picta* has a creamy yellow central band. *Agave angustifolia* has stiff radiating leaves and a prominent terminal spine. *Agave angustifolia* var. *marginata* has a marginal yellow line on the leaves. *Agave filifera* has filamentous hairs along the leaf margins and resembles *Agave angustifolia* in habit. *Agave weberi* called Weber's Century plant has dark green, thick fleshy leaves that reflex with age. The leaf margins are spineless but they terminate in a sharp spine.

Aloe Linn. Asphodelaceae

Aloes

Plants are succulents having extensive pharmaceutical uses and their species diversity is high in Africa. When compared to the Agaves, which they resemble in many features, they are polycarpic. The flowers are borne in spikes or racemes. A number of ornamental species and those with unusual plant forms are grown in gardens of the tropics. They include the following: -

Aloe arborescens
Syn. *Aloe natalensis*

The Candelabra Aloe, Torch Aloe

Plants grow to a height of 2.5m. The leaves are incurved with short marginal spines. The flowers are red-orange and the inflorescence is torch-like.

Aloe aristata

Lace Aloe

Plants have dense rosettes of thick fleshy ovate leaves with pointed tips. Raised white cushiony spots traverse the leaves. The flowers are orange in colour.

Aloe barbadensis

Medicinal Aloe, Kattar-Vazha

Plants have fleshy leaves with soft marginal spines. The leaves are patterned with faint markings. The flowers are yellow and are held clustered on erect, branched inflorescences.

Aloe camperi
Syn. *Aloe eru*

Bedding Aloe

Plants are tufted with a spreading habit. The leaves are white spotted and have reddish brown small marginal spines. The flowers are pink to orange-yellow in colour.

Aloe ciliaris

Climbing Aloe

Plants have hairy leaf-sheath bases and short and stiff marginal spines. Plants are semi-epiphytic and partially shade loving. The flowers are red and yellow.

Aloe dichotoma

The Quiver plant

These are giant succulents with tree-like stems that branch dichotomously and carry rosettes of fleshy leaves terminally. The branches are soft and were used by native Africans to hold their arrows.

Aloe distans

The Jeweled Aloe

The plants are trailing in habit. The leaves are wedge-shaped and coloured bluish green to reddish green. The margins and spines are bright yellow.

Aloe ferox

The Bitter Aloe

The plants are trunk forming and attain heights of up to 4.5m. The leaves are gray-green and become red-tinged with ageing. The margins and lower surface of the leaves have white spines. The flowers are reddish-orange in colour.

Aloe maculata

Syn. *Aloe latifolia*, *Aloe saponaria*

Broadleaf Aloe, Soap Aloe

Plants are herbs attaining heights of up to 1m. The leaves are ovate-lanceolate with arching and tapering tips. The lamina is banded by a transverse pattern of white spots. The margins are edged with reddish brown spines. The flowers are pink and yellow in colour.

Aloe plicatilis

Fan Aloe

Pants are arboreal, attaining heights of up to 4.5m. The strap-like distichously arranged leaves having rounded tips are found towards the tips of branches. The inflorescences carry few small flowers that are bright red in colour.

Aloe speciosa

Tilt-head Aloe

Plants are arboreal attaining heights of up to 4.5m. The leaves are whitish green in colour and become reddish with age. They are held tilted at an angle. The marginal spines are soft and red. The flowers buds are tubular, dark red and open-out to show a greenish white inner surface with red stamens.

Aloe variegata

Ornamental Aloe, Partridge-breasted Aloe

Plants are fleshy succulents with short condensed stems. The leaves are ovate, thick, fleshy and held in a spiral spectacularly marked with transverse bands of light and dark green. The flowers are borne on thick fleshy and green inflorescences and are coral-pink in colour They are pendant.

Cissus cactiformis Vitaceae

Hadjora, Paranda (Mal.)

Plants have stout angular stems that are thick at the nodes. Leaves are present or frequently absent and tendrils occur opposite to leaves. Plants have greenish flowers and are grown as clambering shrubs for their ornamental stems.

Cotyledon orbiculata Crassulaceae

Cotyledons

Plants are sub-shrubs in habit with decumbent cylindrical branches. The leaves are grayish white, thick and fleshy with a flattened spoon like lamina with a narrow base. They are held like a circular stepladder on the stem. Plants are suited for display in containers.

Cotyledon undulata Crassulaceae

Silver-crown Cotyledon

Plants are bushy, attaining a height of up to 50cm. They have thick, fleshy rounded grayish green leaves that have wavy scalloped edges. The flowers are orange.

Crassula arborescens Crassulaceae

Silver Jade plant

Plants grow to a height of up to 2.5m. They are salt-tolerant and are winter-flowering. The leaves are bluish white with pinkish-purple colour along the margins. The flowers are white and have red stamens.

Crassula falcate Crassulaceae
Syn. *Crassula perfoliata*

Paint brush plant, Propeller plant, Sickle plant

Plants are short with bushy to trailing stems and thick, fleshy, grayish green leaves that are sessile, sickle-shaped and turned downward. The flower heads are green and carry small bright red flowers in dense terminal umbels.

Crassula lycopodioides Crassulaceae

Watch-chain Crassula

Plants are low-growing bushy succulents. The stems carry minute leaves packed in 4 rows. The flowers are very small and greenish coloured.

Crassula ovata Crassulaceae

Baby Jade, Jade plant, Japanese Rubber Tree

Plants are short, bushy and carry shiny jade-green fleshy, ovate leaves that become pigmented with a red colouration. They are clustered in whorls at the tips of branches. The flowers are pinkish white in colour.

Dasylirion wheeleri Ruscaceae

Desert Spoon, Bear grass

Plants are unbranched attaining heights of up to 2.5m. Leaves are linear, lanceolate grayish green with marginal hooked spines and are held radiating symmetrically. Terminal spines are absent but the leaf tips are twisted and dry. Inflorescences attain heights of up to 6m. and carry creamy white flowers.

Echeveria derenbergii Crassulaceae

Succulent rose

Plants have thick short stems on which the fleshy leaves are spirally arranged as in a rosette. The length of the leaves gets gradually reduced towards the center. The upper leaf axils carry inflorescences. The flowers are bright red and very small.

Echeveria harmsii Crassulaceae

Painted Lady

Plants are dwarf, bushy in stature and have small linear lanceolate leaves. The flowers are red and are produced in loose upright clusters.

Euphorbia antiquorum Euphorbiaceae

Fleshy Spurge, Malayan Spurge

Plants are branching succulents attaining heights of up to 3.5m. The main stems are 4 to 5 angled and the branches are 3-angled with distinct constrictions at the joints. The leaves are small and rounded.

Euphorbia gymnonota Euphorbiaceae

Bahamas Poinsettia

Plants are hardy evergreens that are salt-tolerant. The leaves are lanceolate and the branches are radiating outwards. The flowers are yellow and have red stamens.

Euphorbia heptagona Euphorbiaceae

Milk barrel Euphorbia

Plants are low growing clump-forming perennials. The stems are cylindrical and many-angled and spiny along the ridges. Leaves are rudimentary, brown in colour and the flowers are also very small and inconspicuous.

Euphorbia ingens Euphorbiaceae

Black bush

Plants are arboreal, attaining heights of up to 9m. The stems are 4-angled and the edges are lined with purplish black tufts of hairs. The leaves are rudimentary. Flowers are very small and found at the tips of stems. The showy fruits are and clustered at stem tips.

Euphorbia leucocephala Euphorbiaceae

Snow bush, Little Christmas flower

Plants are shrubs attaining heights of up to 3m. The leaves are linear and elliptic in outline. Flowers have showy white bracts and are produced profusely in late winter.

Euphorbia lomi, *Euphorbia mili* (See under Flowering Shrubs for the Tropics)

Faucaria tigrina Aizoaceae

Ice plant

Plants have ovate succulent leaves that are covered with shiny projections that glisten like crystals providing camouflage for the plants.

Furcraea striata Agavaceae

Variegated Fourcroya

Plants are Agave-like in habit and attain heights of up to 1.5m. The leaves have band-like variegations of white and light green. The inflorescence stalk grows up to 12m. The flowers are greenish white.

Furcraea marginata Agavaceae

Fourcroya

Plants attain heights of up to 1m and have very short condensed stems. The leaves are held in a rosette and have marginal brown spines that are upturned.

Haworthia limifolia Asphodelaceae

Haworthia

Plants are low-growing hardy succulents having fleshy leaves arranged in a rosette. Their leaves tend to cover containers in which they are grown entirely. Their surfaces have felt-like soft white hairs.

Haworthia margaritifera Asphodelaceae

Pearl plant

Plants resemble the above species in habit. The leaves are spotted white with cushiony projections.

Haworthia truncata Asphodeliaceae

Window plant

Plants have distichously arranged leaves that have a warty surface texture and an oval outline. The flat tips of the leaves are translucent.

Kalanchoe beharensis Crassulaceae

Felt bush, Velvet leaf

Plants have partially folded leaves that are covered with silvery gray or bronze hairs and have wavy margins that are spiny. Flowers are small and saucer shaped and are produced in many flowered panicles in winter.

Kalanchoe blossfeldiana Crassulaceae
Syn. *Kalanchoe globulifera* var. *coccinea*

Christmas Kalanchoe, Calanchoe, Flaming Katy

Plants are short-day flowered and are popular in varying climatic zones throughout the world. The leaves are elliptic-ovate and have broadly crenate margins. Though red flowered plants are more popular, yellow, pink, orange and white flowered cultivars are also grown.

Kalanchoe crenatum Crassulaceae
Syn. *Bryophyllum crenatum*

Scallop-leaf Bryophyllum

Plants are entirely greenish white in colour. The leaves are broadly ovate and the lamina is held at an angle and chanelled towards the tip. The leaf margins are crenulate. Plantlets of different sizes that appear in leaf notches add to the unusual appearance of plants.

Kalanchoe daigremontianum Crassulaceae

Palm tree Bryophyllum

Plants are slender stemmed perennials. The cylindrical shoots of up to 1cm thickness are closely marked by leaf scars. The leaves are cylindrical, bluish green and faintly marked with brownish black spots. The leaf margins carry minute plantlets that are shed one after the other. The bare stems with the upper whorl of leaves give plants the appearance of miniature palm trees.

Kalanchoe delagoensis Crassulaceae
Syn. *Bryophyllum tubiflorum*

Mother-of -thousands

Plants are hardy herbs having elongated dark green leaves arranged in whorls around the stem. A large number of plantlets arising from the leaf margins and remaining pendant have given rise to the common name of "mother -of-thousands". The flowers are tubular, pendant and pink in colour.

Kalanchoe grandiflora Crassulaceae

Malakalli

Plants are hardy herbs having light green ovate leaves that are surfaced with a white powdery waxy coating. The margins are broadly crenulate. The flowers are tubular wth yellow pointed free lobes opening outwards. They are held erect terminally on corymbose clusters.

Kalanchoe thyrsiflora Crassulaceae

Blade leaf Kalanchoe

Plants have characteristically valvate sessile leaves with rounded tips that are imbicately arranged and appear to be folded. The leaf margins are reddish purple in colour. The flowers are yellow, bell shaped and held on thyrsiform inflorescences.

Kalanchoe pinnata Crassulaceae
Syn. *Bryophyllum pinnatum*

Air plant, Chandelier plant

Plants are bushy and small leaved. The leaves are pale greenish white and their margins are broadly crenulate.

Kalanchoe marmorata Crassulaceae

Pen Wiper plant

Plants have obovate fleshy leaves that are notched along the margins. The lamina has attractive brown markings on both surfaces.

Kalanchoe tomentosa Crassulaceae

Panda plant, Pussy-ears

Plants have bushy branching stems. The fleshy leaves are spoon shaped and are entirely covered by white velvety felt-like hairs. The broadly toothed leaf margins are edged brown.

Lampranthus aureus Aizoaceae
Syn. *Mesembryanthemum aureum*

Trailing Ice plant

Plants are mat forming perennials having 3-angled small fleshy green stems. The flowers are red-orange to dark brown in colour with yellow stamens.

Lithops lesliei Aizoaceae

Stone plants, Living stones

These are tufted in habit, with the plant body coloured grayish green and shaped like stones. The shoots are obconical with the apex appearing to be flat or truncated. They are also marked with gray and white spots to mimic weathered rocks and stones amongst which they are found in their native habitats.

Beaucarnia recurvata Ruscaceae
Syn. *Nolina recurvata*

Nolina

Plants have slender cylindrical rough-surfaced stems that enlarge at the base with age, forming a curved club shaped structure. The leaves are crowded towards the tips of branches. They are slender, grass like, recurving and about 2cm wide and more than a meter long.

Pachyphytum oviferum Crassulaceae

Moonstones

Plants are bushy succulents attaining heights of up to 15cm. The leaves are ovate, bulbous, grayish green and are tipped pink. Flowers are red.

Pedilanthus tithymeloides Euphorbiaceae

Slipper plant, Slipper spurge, Bird cactus

Plants are sub-shrubs attaining heights of up to 1m. The leaves are leathery, dark green and are held on rounded stems that curve irregularly. In *Pedilanthus tithymeloides* var. *variegatus,* the leaves are white to light green and tinged pink. In *Pedilanthus tithymeloides* var. nanus, the leaves are smaller, dark green and distichously arranged on darker green stems.

Portulacaria afra Portulacaceae

Elephant bush

Plants are partially shade loving shrubs attaining heights of up to 3m. The young shoots are fleshy and red tinged. The leaves are small, broadly ovate to round and thick. Plants can be grown in rockeries and in pots.

Sansevieria arborescens Ruscaceae

Snake-plant tree, Tree Sansevieria

Plants are rhizomatous, producing erect pseudostems formed by sheathing leaf bases. The leaves are fleshy, fibrous and the blades terminate in sharp spines. The lamina is characteristically patterned as in the other species of the genus.

Sansevieria aubrytiana Ruscaceae

Snake plant

Plants have clusters of erect leaves that arise from the rhizome. The surface of the lamina is patterned with creamy white mottling resembling the pattern of snake's skin. The flowers are creamy white and are produced on erect spikes.

Sansevieria cylindrica Ruscaceae

St. George's Lance

Plants have cylindrical lance-like leaves with broadly pointed tips. They arise from the rhizomes and are initially upright and become decumbent with ageing.

Sansevieria kirki Ruscaceae

Snake plant

Plants are low growing and have erect spreading clusters of shoots. The leaves are dark grayish green, shiny and mottled in creamy white. The leaf margins are wavy. The rhizomes are reddish orange.

Sansevieria trifasciata Ruscaceae

Bowstring hemp

Plants are perennial rhizomatous herbs that spread freely. The leaves are dark green, lanceolate and marked with faint light green transverse markings. They are grown indoors and outdoors as foliage plants. *Sansevieria trifasciata* var. *vareigata* cv. 'Laurentii' has darker green markings on the leaves and a golden yellow marginal band. In cv. 'Hahnii', plants are low growing and the leaves are small and are held in a rosette.

Sedum brevifolium Crassulaceae

Trailing Sedum

Plants have succulent trailing fleshy stems that carry pendant sickle shaped fleshy leaves. They are ideal for hanging pots.

Sedum lineare Crassulaceae

Vareigated Sedum

Plants are bushy and semi trailing in habit. The shoots are semi erect and carry very small, linear lanceolate leaves that are variegated in cream and white. The flowers are yellow. Plants are suited for well-lighted indoor areas.

Sedum morganianum Crassulaceae

Donkeys Tail

Plants have trailing stems that carry closely spaced ovate fleshy leaves. They are suited for hanging pots.

Sedum nutans Crassulaceae

Green-rose Sedum

Plants have oblong ovate fleshy leaves with flattened surfaces. The leaves are held in terminal clusters on shoots and appear as attractive rosettes.

Sedum pachyphyllum Crassulaceae

Jade Beads

Plants have ovate ball like fleshy leaves that cluster all along the shoots like green beads. They are ideal for pot growing indoors.

Sedum rubrotinctum Crassulaceae
Syn. *Sedum guatemalensis*

Stonecrop, Orpine

Plants are mat forming succulent herbs with fleshy club shaped leaves that are red tinged at the tip. The flowers are small, yellow, star shaped and are found in short inflorescences.

Stapelia gigantea Apocynaceae
Syn. *Stapelia nobilis*

Giant Starfish flower

Plants are low growing succulent stemmed herbs attaining heights of up to 30cm. The rudimentary leaves produced are shed quickly. The stems are produced in clusters and have four ridges that carry upright hooks. The flowers are solitary and have a star shaped corolla of 30cm width with spreading lobes and pointed tips. The corolla is red with yellow transverse corrugations that are lined with red hairs. They emit a foul odour.

Stapelia grandiflora Apocynaceae
Syn. *Stapelia ambigua*

Starfish flower, Carrion flower

Plants are low growing succulent stemmed herbs resembling *Stapelia gigantea*. The flowers are smaller, up to 15cm in width, dark red with purple hairs.

Welwitchia mirabilis Welwitschiaceae

Tumboa, Welwitschia

These plants are native to the coastal sandy deserts of Namibia. After the seed leaves, they produce two leaves only that continue to grow from the base throughout their life for more than 500 to 600 years, with their tips alone drying on ageing and disintegrating with time. They are dioecious and produce salmon-pink to dark pink cones.

Yucca aloifolia Agavaceae
Syn. *Yucca gloriosa, Yucca yucatana*

Spanish Bayonet, Dagger plant

Plants are tall growing evergreens attaining heights of up to 4.5m. The leaves are thin and lanceolate with terminal spines and are held closely arranged around the stems. Old and dried leaves persist hanging downwards, covering the basal portion of the stem. The stems become decumbent with age and develop shoots profusely, resulting in large clumps. The cultivar 'Marginata' has a creamy white colour along the margins.

Yucca elephantipes *Agavaceae*

Stick Yucca

Plants are similar in habit to *Yucca aloifolia.* The leaves are broadly strap like with pointed tips. Plants are used extensively for landscaping both indoors and outdoors.

Yucca rostrata Agavaceae

Beaked Yucca

Plants are arboreal evergreens similar to the above two species in habit. The leaves are linear, with white lined margins and are held radiating all around the terminal ends of branches. The seeds are beak shaped.

CACTI

Cacti are members of the family **Cactaceae**. They are dicotyledonous perennial stem succulents and are mostly spiny and leafless. The presence of areoles or cushion like pads from which spines, hairs, branches or flowers arise distinguish them from other succulents. In their native habitats, cacti flower periodically and may be day-flowering or night-flowering. The flowers are white to dark red, purple or yellow and are short-lived. The flowers that open at night are white or pale coloured and scented and remain closed during daytime. The day bloomers are brightly coloured and have a characteristic metallic luster. To enable regular flowering, cacti are to be grown under moisture-stressed and well lighted situations They require coarse growing media with pH in the alkaline range. Different genera of cacti can be readily grafted. Grafts tend to flower more quickly than parental species and result in unusual plant forms. They are generally short-lived.

Popularly grown ***Ornamental Cacti*** of gardens include the following:

Aporocactus flagelliformis

Rat-tail cactus

Plants have cylindrical creeping stems that are closely ribbed and covered with small brown spines. The flowers are dark red in colour.

Ariocarpus fissuratus
Syn. *Anhalonium fissuratum*

Horned star cactus

The plant body is surfaced by projections and fissures and is shaped like a shriveled star. The flowers are pink. In *Ariocarpus retusus*, the plant body is globular, appearing compressed with a rosette of grayish green erect tubercles held like leaves. The flowers are yellow.

Astrophytum asterias

Wooly star cactus

Plants have rounded stems. The ribs are hairy and spineless. The flowers are yellow. *Astrophytum ornatum*, the spiny star cactus has globular twisted ribs with long spines.

Cactus melocactoides

Wooly cap cactus

Plants produce a wooly cap-like cephalium terminally when the main stem stops growth. The cephalium has brown bristles and grows every year. The flowers are pink in colour.

Cephalocereus senilis

Old man cactus

Plants attain heights of upto 12m and have a columnar body that is closely ribbed and covered with long grey hairs. The flowers are pink.

Cephalocereus cristatus

Crested old man cactus

Plants are crested and tall growing and the body is covered with fine pure white hairs. The flowers are rose pink.

Cereus peruvianus

Curiosity plant, Peruvian Apple Cactus

The plant body is bluish green and has unusual irregular growth that seems monstrous. Prominent spines cover plants. In a dwarf-statured cultivar, the spines are thinner and brown in colour.

Chamaecereus sylvestrii

Peanut cactus

Plants have short clusters of cylindrical stems-often spiraling- that are covered with thin white spines. The flowers are orange-red. A variety with crested stems is also grown.

Cleistocactus baumanni

Orange-torch cactus

Plants have thin columnar stems that branch at the base. White, yellow or brown spines cover them. The flowers are orange-red.

Cleistocactus straussi

Silver-torch cactus

Plants produce thin columnar stems in clusters. They are covered by short yellowish white spines. The flowers are dark red in colour.

Coryphantha echinus

Ball cactus

Plants have solitary globular stems that attain diameters of up to 5.0 cm. They have white spines that are appressed radially. The flowers are yellow.

Cryptocereus anthonyanus

The Ribbed cactus

The plant body is flattened into rib-like lobes that are often prostrate giving plants a creeping habit. Spines are absent.

Echinocactus grusonii

Golden Barrel cactus, Mother-in -law's Chair

The plant body grows into a large globular structure, attaining a diameter of up to 1m. One or two smaller globular stems are occasionally seen at the base. The plant body has 27 to 37 prominent ribs that are cushioned by golden yellow spines. Flowers are borne at the apex of the stem. *Echinocactus grusonii* var. *cristatus* is a crested form.

Echinocereus dasycanthus

Rainbow cactus

The plant body is small and cylindrical and has radial spines that change colour with age. They are pinkish when young and slowly turn white and later become grey. The variation in colour seen in spines of different ages has resulted in its popular name.

Echinocereus longispinus has long creamy white spines that are tipped red.

Echinopsis multiplex

Hedgehog cactus

Plants are low growing with short cylindrical or globular stems on which spines are arranged radially.

Epiphyllum oxypetalum

Night-blooming cactus, Nishagandhi

Plants are climbing epiphytes having flattened leaf-like stems. They are spiny when young. The flowers are large, fragrant, sometimes day-blooming and coloured white or light orange.

Espostoa lanata

Peruvian Old man cactus

Plants have tall, columnar, occasionally branching stems that are covered with cottony white hairs carrying spines. The flowers are pink in colour. *Espostoa plumosa* has smaller spines and a dense cover of white hairs.

Ferocactus horridus

Hooked-spine cactus

Plants are globular with 8 to 13 broad ribs. The radial spines number 8 to 10. One is long, flattened and hooked while the others are straight and red in colour. *Ferocactus latispinus* has large globular and compressed stems having 15 to 23 broad ribs. The radial spines numbering 8 to 10 are pink to white and the single central spine is flat and hooked, and the rest are thick and dark pink in colour. *Ferocactus pringlei* has clump-forming stems that are initially globose and later become cylindrical attaining heights up to 3m. There are 16 to 18 compressed ribs that carry sharply curved red spines.

Gymnocalycium mihanovichii

Chin cactus

Plants have a grayish green stem. The ribs are cut into chin-like warts with red bands. There are 8 acute ribs with spreading radial spines that are creamy yellow in colour. *Gymnocalycium multiflorum* is low growing with globular or columnar stems having 10 to 15ribs that carry flat radial spines. *Gymnocalycium quehlianum* resembles a flattened globe carrying 8 to 13 ribs having radial recurved spines. *Gymnocalycium saglione* has large globular stems having 13 to 32 ribs that are broad, carrying curved brown to black spines.

Hylocereus undatus

The Climbing cactus

Plants are epiphytic climbers having 3-angled stems. The flowers are large, 25 to 30 cm long and are white in colour. They are night blooming and open for one night only.

Mamillaria spp

Powder-puff cactus

Plants are small, rounded or broadly cylindrical in shape and the stems are covered with white hairs. *Mamillaria camptotricha,* commonly called the Bird's nest cactus has a plant body made up of small globular clusters bearing twisted yellow spines. The 'Thimble cactus'- *Mamillaria fragilis* forms thimble-like branches at the apex that break off easily. *Mamillaria hahniana,* the 'Old Lady cactus' has a flattened globose body that is covered with long white hairs and spines.

Melocactus meolcactoides

Wooly Cap cactus

Plants are small, conical and have 10 broad ribs having white radial spines. They have a cap-like structure called cephalium that is made up of white wooly hairs and brown bristles and is devoid of spines. The flowers are pink in colour.

Notocactus haselbergii

Miniature Ball cactus

Plants are very small, globular, 7 to 8 cm in diameter and have soft silvery white spines with yellow tips. The flowers are coloured saffron-red. *Notocactus leninhausii* attain heights of up to 1m and have cylindrical columnar stems. The yellow spines cover the closely spaced ribs. The flowers are lemon yellow. In *Notocactus scopa,* the spines are long brown and needle-like. An ornamental cristate variety with flattened and coiling stems covered with white hairs is also grown.

Opuntia clavarioides

Sea Coral cactus

Plants have short stems that are finger-like or fasciated and resemble sea corals having similar shapes.

Opuntia microdasys

Bunny-eared cactus

These have oval pad-like stems that arise on older stems. These resemble ears. Minute brown velvety tufts of bristles that darken with age cover the aereoles. These detach easily and stick readily on to skin and clothing and are difficult to remove. Several other species are found to have ornamental value and include the following: -

Opuntia tetracantha is bushy and has cylindrical joints with needle-like spines. The flowers are green tinged with purple. *Opuntia tomentosa* is arboreal, attaining heights of up to 3 to 5m. The spines are green in colour and the flowers are yellow.

Opuntia tunicata has whorled branches, is short in stature and the plants are entirely covered with barbed spines that are sheathed in white. The flowers are yellow. *Opuntia vulgaris* is fleshy and more or less spineless. The joints of the oval stem pads are flattened and often pattered green and white. Viviparous germination is found in mature unripe fruits.

Oreocereus celsianus

Red-spine cactus

Plants are clump forming and attain heights of up to 1m. The branches are prostrate at the time of emergence and curve upwards and attain a thickness of up to 8m. The ribs number 10 and the areoles have long hairs and red spines. The flowers are dark red. Other ornamental species having columnar stems entirely covered by silvery to creamy white hairs are *Opuntia maxima*, *Opuntia ritterii* and *Opuntia trollii*.

Parodia aureispina

Tom-thumb cactus

Plants are small, rounded and are coloured bluish green. The ribs are spirally twisted. The spines are white and yellow coloured at the center. The flowers are golden yellow in colour. in *Parodia massii*, the stems are greenish yellow and rounded in shape. The ribs are spirally twisted and carry radial spines on knob-like structures. The spines are white and the central spine is hooked.

Pereskia aculeata

Rose cactus, Leafy cactus

Plants are vine-like to bushy in habit and have woody stems. The spines are brown in colour. The flowers are creamy white and scented.

Rebutia fiebrigii

Rebutia

Plants are low growing and have a clustering habit producing offsets that root easily. Tubercles on the stems are spiraled and carry white spines. The flowers are orange-red in colour.

Rebutia minuscula

Red Crown cactus

Plants are globular, clump forming and are covered by creamy-pink spines. At flowering, plants are covered by a mop of dark red flowers.

Rhipsalidopsis gaertneri

Easter cactus

Plants are bushy to trailing in habit and have flattened spatulate stem segments with broadly crenulate or notched margins. The flowers are produced at stem tips and have many pointed bright red petals.

Rhipsalidopsis rosea

Rose Easter cactus

Plants have small stems and are low growing and mat-like. The stem segments are bluish green and they produce rose pink flowers in profusion.

Rhipsalis spp

Rhipsalis, Wickerwork cacti

Plants are epiphytes having branched stems that are cylindrical, angular or flattened. The wooly or hairy areoles are irregularly scattered and mostly spineless. A number of species are grown in gardens as specimen plants and include *Rhipsalis baccifera* the Mistletoe cactus and *Rhipsalis paradoxa* the Chain or Link plant.

Selenicereus grandiflorus

Vanilla cactus

Plants are epiphytic climbing shrubs with stems attaining a diameter of 2.5cm. Flowers open at night. They are 30cm long, white on the inside and salmon pink on the outside and vanilla scented. *Selenicereus urbanicus* has angled stem joints and aerial roots. The flowers are light brown coloured on the outside and white within.

Trichocereus spp.

Columnar cacti, Trichocereus

Plants have columnar stems with branched ribs. The flowers are large, white and scented. The floral tube has scales and hairs. A number of species are grown in gardens. The genus is now placed in *Echinopsis,* the genus of Sea Urchin cacti or Easter Lily cacti.

Zygocactus truncatus

Crab cactus, Thanksgiving cactus

Plants are epiphytic and have flattened dichotomously branched stems with serrate branches.

contd. colour plates

Adenium obesum

Agave weberi

Euphorbia antiquorum

Euphorbia mili 'Minibell'

Euphorbia x lomi

Kalanchoe blossfeldiana

Chapter 7

ORNAMENTAL BROMELIADS

Terrestrial or epiphytic members of the family **Bromeliaceae** that are grown in gardens for their unusually coloured and patterned leaves or showy inflorescences and fruits are referred to as ornamental '**Bromeliads**'. Being low growing, they are ideally displayed in containers or as epiphytes. They are sun loving, suckering in habit and have xerophytic adaptations. The leaves are tough and fibrous often with spiny margins and sheathing leaf bases that hold water. A grayish-white waxy or powdery coating often covers the lamina. Certain bromeliads have a water collecting depression at the center. Inflorescences are sunken, erect or reclining, with condensed or long and much branched stalks carrying colourful bracts and usually, less showy flowers. In many, the terminal leaves become brightly coloured at flowering. These and the floral bracts remain showy even after the flowers are shed.

Ornamental **bromeliads** popularly grown in gardens include the following:

Aechmea blanchetiana

Copper leaf bromeliad

Plants have strap like leaves that develop streaks of coppery red colour on the upper surface in bright sunlight.

Aechmea burle-marxii

Burle-Marx bromeliad

An ornamental bromeliad named after the famous Brazilian landscape architect Burle Marx, this has leaves that are transversely banded in green and gray. The flowers are small, yellow and tubular and are held in the axils of pink bracts.

Aechmea chantini

Amazon Zebra plant, Queen of Bromeliads

Plants are semi epiphytic herbs. The leaves are transversely banded with white and dark green lines. The inflorescence is branched and carries heliconia-like yellow flowers in the axils of red bracts. A number of cultivars with varying bract colours and leaf patterns are grown in gardens.

Aechmea fasciata

Urn plant, Silver vase

These epiphytes have an overall resemblance to an urn. The leaves are grayish white and are gracefully reflexed. The central inflorescence stalk has bracts that are coloured bright pink. The flowers are pink and blue coloured. They are shed quickly but the bracts that subtend them have a greater longevity.

Aechmea fulgens

Coral Berry

Plants are semi epiphytic low growing herbs. The leaves are spineless and arranged in reflexing rosettes. The central leafy bracts that subtend the branched inflorescences are red. The flowers are blue-violet in *Aechmea fulgens* var. *discolor* and the leaves are have longitudinal lines or bands of white and green.

Aechmea mariae-reginae

Queen Mary bromeliad

Plants are dioecious. The leaves are found in rosettes of up to 1m widths. The male and female inflorescences are cylindrical condensed spikes carrying small white flowers with bright pink bracts. The female inflorescence is shorter and more compact than the male Inflorescence.

Ananas bracteatus var. *tricolor*

Variegated pineapple, Ornamental Pineapple plant

Plants have profuse suckering. The leaves have greenish yellow margins with an orange-red to dark red central band. The inflorescences are carried terminally on long stalks. The fruits are small, inedible, coloured yellow and red with pink bracts.

Ananas comosus var. *vareigatus* has narrow and lighter green leaves than the cultivated species. A creamy yellow to light pink band runs along the margins of the leaves. The leaf margins are spiny. The fruit is pink and green and is topped by a bright pink crown and is pineapple-like.

Bllbergia nutans

Angel's tears, Queen's tears

Plants are bushy, and have flaccid ribbon-like leaves that look untidy. The inflorescences are elongated, slender and pendant and are covered by orange-red bracts that are very showy. In cv. Fantasia, the leaves are dull green in colour and heavily spotted with yellow and the shoots are erect and the leaf tips are reflexed. Under bright sunlight plants become entirely shaded with a rose-pink colouration.

Billbergia pyramidalis
Syn. *Billbergia bicolor / Bromelia pyramidalis*

Summer Torch

Plants are low growing and have light green coloured leaf rosettes. The inflorescence is showy and has an erect, club shaped red stalk and bracts. The petals are red to purple.

Cryptanthus bivittatus
Syn. *Cryptanthus zonatus*

Earth Star

Plants are low growing ground hugging in habit. The leaves are green to brown coloured with white to creamy yellow longitudinal or zigzag transverse bands. The leaf margins are wavy. A number of cultivars with well-marked leaves are grown.

Cryptanthes bromelioides

Rainbow Star

Plants are low growing and few leaved. The leaves have prominent green and creamy white lines. In bright light the white bands and lines become dark pink along the edges. The leaf margins are undulating.

Guzmania lingulata

Scarlet Star, Quiche

Plants are epiphytic herbs having long ribbon-like leaves that are soft and shiny with smooth margins held in a rosette. At the time of flowering, a group of bright red leaf like bracts arises from the center. These are more long lived than the small white flowers. In cv. 'Cardinalis', the inflorescence resembles a torch or a fountainhead.

Guzmania sanguinea

Scarlet Star

Plants are epiphytic and have a leaf spread of up to 0.5m. The glabrous green leaves become bright red or streaked with vermilion streaks at flowering. The floral bracts are red, green and yellow. Plants are spectacular when in flower.

Guzmania wittmackii

Vareigated Scarlet Star

Plants have slender arching leaves that are variegated with white lines. The inflorescence is erect attaining a height of up to 1m. It carries white flowers subtended by green and red bracts.

Neoregelia carolinae

Blushing Bromeliad

Plants are similar in habit to Billbergias. The leaves are strikingly coloured with an olive-green colour and have a central creamy yellow band. Before flowering the central rosette of leaves become partially or completely red. The inflorescence is condensed and has small blue or white flowers. *Neoregelia concentrica* has inflorescences that are almost sunken at the center of plants. The central leaves turn violet at flowering.

Neoregelia cruenta

Painted Finger nails

Plants have attractive olive green to bright green leaves with pink to red tips. The leaf margins are finely toothed and slightly concave towards the tip. The inflorescences appear partially sunken and are violet in colour with light purple flowers. A number of cultivars with variously spotted and lined leaves are grown in tropical gardens.

Nidularium fulgens

Red-hearted Bromeliad

Plants have bright green leaves that are faintly mottled with yellow. The leaf margins are spiny. The inflorescence has a condensed stem covered by red bracts. The flowers are violet-blue in colour.

Nidularium innocentii

Bird's Nest Bromeliad

Plants have a rosette of dark green leaves that are faintly shaded with a purplish maroon colour. The lower surface of the leaves is dark red. The bracts on the condensed inflorescence axis are orange-red and the flowers are white in colour.

Pitcairnia angustifolia

Red-head Bromeliad

Plants have a bushy habit and grass like leaves. The inflorescence is upright; much branched and carries small red flowers.

Tillandsia capitata

Purple Torch

Plants are semi epiphytic herbs having a lateral spread of up to 50cm. The sword shaped leaves turn reddish at flowering. The flowers are purple and the bracts are red.

Tillandsia cyanea

Blue Torch

Plants are bushy with narrow grass like leaves that arch outwards. The terminal inflorescence is erect, broadly ovate and consists of distichously overlapping green bracts that turn dark pink before flowering. The violet-blue flowers are exerted from the bracts before opening. The bracts remain showy for a few weeks after the flowers are shed.

Tillandsia dyeriana has long pendant inflorescences carrying distichously arranged bracts resembling Heliconias. They are golden yellow initially and turn saffron-red at full bloom.

Tillandsia ionantha

Silver Urn

Plants are epiphytic, having fleshy tapering leaves with arching tips. They are covered with short white scales. The leaves at the center that subtend the flower heads become red in colour. The flowers are violet in colour.

Tillandsia usneoides

Syn. *Tillandsia filiformis*

Spanish Moss

Plants are small epiphytes having semi-terete bluish green leaves that are covered by grey hairy scales. Long filamentous stolons produced by plants hang downwards from their perches on trees. The flowers are bluish green.

Vriesea splendens

Syn. *Vriesea ensiformis / Vriesea conferta*

Flaming Sword

Plants have a rosette of leaves that are banded with brown and green colours. The inflorescence is held upright and the upper half carries orange-red bracts. These subtend tubular yellow flowers. The orange-red bracts remain showy while the flowers are shed quickly.

contd. colour plates

Aechmea fasciata

Ananas bracteatus var. *Tricolor*

Ananas comosus var. *Variegatus*

Ananas nanus

Billbergia nutans

Neoregelia cruenta

Chapter 8

FERNS AND FERN ALLIES

Pteridophytes such as ferns belonging to several families, selaginellas popularly called 'club mosses' belonging to **Selaginellaceae** and the Lycopods which resemble both these, belonging to **Lycopodiaceae** are shade loving plants adapted to cool and moist or hot and humid ecosystems of tropical to temperate climates. They are mostly epiphytic or lithophytic and are grown as foliage ornamentals in a fernery or in shaded areas in gardens. Many are favoured indoor plants.

Popularly grown **ferns, lycopods** and **selaginellas** of gardens are listed below:

Adiantum capillus-veneris Adiantaceae

Common Maidenhair fern

Plants are shade loving bushy lithophytes and have delicate looking thin black stems carrying light green oblique-ovate fronds. They are pot grown as ornamental foliage plants.

Adiantum decorum Adiantaceae
Syn. *Adiantum raddianum*

Maidenhair fern

Plants are bushy with thin, wiry shiny black stems resembling hairs. The fronds are oblique ovate and finely crenulate. Plants and leaves are widely used decoratively.

Asplenium bulbiferum Aspleniaceae

Hen and Chicken fern

Plants are bushy in stature and have finely divided pinnate fronds. The mature fronds carry spores that can be detached and used for raising new plants.

Asplenium nidus Aspleniaceae
Syn. *Asplenium nidus-avis*

Bird's Nest fern

Plants have broad, ovate-lanceolate entire leaves with a dark brown prominent midrib. The fronds are held in an open cluster resembling a nest. A number of species with finely divided fronds are also grown of which *Asplenium trichomanes* is a rare epiphyte of the lower Himalayas.

Athyrium nipponicum Athyriaceae

Japanese painted fern

Plants are attractive foliage plants. The young leaves and midrib are reddish purple in colour. *Athyrium drepenopterum* is a well-distributed species of the northern hill zones.

Blechnum gibbum Blechnaceae

Palm leaf fern

Plants are bushy with pinnately divided fronds that attain lengths of up to 1m. The linear pinnae have wavy margins and sharply pointed tips. A trunk like black stem becomes visible as plants age.

Cyathea spp Cyatheaceae

Tree ferns

These are ferns found in the tropics and the southern hemisphere. They are large in size when compared to other herbaceous pteridophytes and have palm like trunks carrying a terminal crown of pinnate leaves that unfurl characteristically. *Cyathea brunoniana* Hook. is well distributed in the western Himalayas. *Cyathea capensis* L. and *Cyathea dregei* Kunze. are indigenous to Uganda. *Cyathea medullaris* G. Forst. the 'Mamaku' of New Zealand is the largest among the tree ferns grown, attaining heights of up to 15m. The stout trunks are unbranched and swollen in this species and the leaf stalks are scaly.

Cyrtomium falcatum Dryopteridaceae

Holly fern

Plants are bushy in habit and the fronds are held on upright stalks having outlines resembling holly leaves.

Davallia canariensis Davalliaceae

Deer's foot fern, Hare's foot fern

Plants have creeping rhizomatous stems covered with dark brown to black felt like hairs. The fronds are triangular in outline and are finely cut and arise directly from the rhizomes, held on upright stalks.

Dicksonia squarrosa Dicksonoiaceae

Dicksonia fern, New Zealand tree fern

Plants have a palm like habit with truncate stems and a crown of pinnate leaves. The leaf stalks are hairy.

Drynaria quercifolia Polypodiaceae

Oak Leaf fern, Basket fern

Plants are hardy epiphytes having fleshy rhizomatous stems that are covered by dry frond like coverings. The leaves are pinnately lobed and are held erect.

Dryopteris marginalis Dryopteridaceae

Heather wood fern, Marginal Shield fern, Wood fern

Plants are bushy terrestrials of rocky terrain. The fronds are broadly triangular, acuminate and are held in a crown. The bipinnate lobes of fronds are ovate to oblong and entire or partially adnate.

Lygodium palmatum Schizaeaceae

Climbing fern, Creeping fern

Plants are shade loving and have thin wiry stems that creep or climb through the under growth of forests. They are widespread in the Western Ghats. The leaves are attractive, palmate and dark green in colour.

Nephrolepis spp. Oleandraceae

Sword ferns

Plants are shade loving, bushy in habit and have arching fronds. They are popular indoor plants of the humid tropics and their fronds are used decoratively. *Nephrolepis acuminata* syn. *Nephrolepis davallioides* called 'the Ladder fern' has narrow acute-tipped pinnae held distichously. *Nephrolepis duffi* has small rounded pinnae. In *Nephrolepis exaltata* the pinnae, held at right angles in two rows have bifid tips and wavy margins. A number of cultivars of this species with curled, ruffled or erect fronds or multifid pinnae are grown in gardens.

Osmunda regalis Osmundaceae

Flowering fern, Royal fern

These are clump-forming ferns adapted to open marshy situations of temperate regions. They attain heights of up to 1.5m. The fronds are feathery crested or plain. The upper part of fertile fronds appears like inflorescences. The root fibers are used as a medium for growing nursery plants.

Pellaea rotundifolia Adiantaceae

Button fern

Plants are bushy and have a spreading habit. The fronds attain lengths of up to 30cm and are held radiating outwards. The small ovate pinnae are held on either sides of the main axis in opposite rows.

Platycerium bifurcatum Polypodiaceae

Stag horn fern

Plants are epiphytes having characteristically lobed fronds that resemble antlers of stags. They are usually grown in containers in gardens or in a fernery and are allowed to cover the outer surface of the container entirely.

Polypodium aurium Polypodiaceae

Hare's foot fern, Polypody

Plants have thick creeping rhizomatous stems that are covered by maroon-red scales. Fronds arise directly from the stems and are pinnate leaved. The pinnae held on either side, arch gracefully.

Polystichum acrostichioides Dryopteridaceae

Rock fern

Plants are hardy and have multifid fronds. Plants are grown indoors and the leaves have decorative use.

Pteris cretica Pteridaceae

Cretin Brake

Plants are clump forming and short in stature. The fronds are held on erect stalks that eventually become decumbent. They have finger-like short or elongated segments numbering a few to nine. Varieties with frilled, crinkled or variegated segments are grown.

Lycopodium spp Lycopodiaceae

Club mosses, Lycopods

Plants have fronds resembling those of selaginellas and erect or trailing fern-like stems. They are displayed along with ferns in ferneries. A number of species having ornamental or rarity value are grown in gardens and include *Lycopodium annotinum*, *Lycopodium clavatum* called Vegetable Sulphur or Wolf's Claw and *Lycopodium taxifolium*.

Selaginella delicatula Selaginellaceae

White-leaf Club moss

Plants are bushy and have thin and narrowly lobed leaves that have a white line along the mid rib of fronds.

Selaginella emmeliana Selaginellaceae

Cut leaf Selaginella / Cut leaf club moss

Plants are bushy and have a thick cluster of finely segmented leaves. The shoots are decumbent and root readily at the nodes. The leaves are used decoratively.

Selaginella kraussiana Selaginellaceae

Spreading Club moss

Plants have thin shoots that are spreading in habit. The growing leaf tips appear bifid and are light green.

Selaginella laevigata Selaginellaceae

Blue-green Club moss

Plants have flattened fronds that are coppery when young and attain blue-green tints in dense shade. The shoots tend to creep and clamber and plants can be ideal ground covers for humid shady corners of a fernery.

Selaginella leptophylla Selaginellaceae

Resurrection plant

Plants are native to desert regions. The leaves roll up and turn into a tight ball and attain a brown colour during the dry period. When water becomes available they open out and become green.

contd. colour plates

Davallia canariensis

'Maiden hair fern'

Nephrolepis acuminata

Nephrolepis duffii

Nephrolepis exaltata

Platycerium bifurcatum

Asparagus meyeri

Asparagus plumosus

Pilea microphylla

Selaginella emmeliana

Selaginella laevigata

Selaginella delicatula

Chapter 9

ORNAMENTAL GRASSES BAMBOOS, SEDGES AND RUSHES

Grasses, bamboos, sedges and rushes are plants that are being used increasingly in xeri-scaping and in marsh gardens for getting the benefits of their unusual foliage textures and colours. Their hardiness, stature and canopy frame make them more suited for specimen planting in gardens and landscapes.

GRASSES

Grasses are the most successful of monocots belonging to the family **Poaceae** that are found the world over. Low growing perennial grasses are used to make attractive lawns in gardens and to provide a desirable turf cover for various situations. A number of species belonging to various genera differing in leaf colour, shape and growth pattern are used for this purpose. Mixtures of different varieties of grasses are also used for making turf having desirable blends of textures and hardiness. Tall bushy grasses having a mound-like or fountain–like habit and fine to velvety surfaces are grown for their textural and architectural features in landscaping. Many are grown as specimen plants for their display value at flowering. In selecting grasses for gardens, their potentially invasive nature is to be taken into account and seed dispersal controlled by timely cutting.

Grass species that thrive in the hot and humid tropics are called warm-season grasses and include Bermuda grass, Carpet grass and St. Augustine grass. Cool season grasses that are adapted to temperate regions and thrive at temperatures of 10 to 20°C include Blue grass, Fescues and Zoysia.

Popularly grown **ornamental grasses** include the following:

Agrostis stolonifera

Creeping Bent grass

Plants are tall, bushy and hardy and have creeping stolons that tend to smother other grass species and weeds. They were popularly used for turfing golf courses before the coming of newer grass varieties. A number of species are native to different parts of the Northern hemisphere.

Anthoxanthum odoratum

Sweet Vernal grass, Vanilla grass

These are perennial grasses that have a sweet vanilla-like aroma. The leaves are used in native medicine. It's pollen cause hay fever.

Arundinaria spp.

Bamboo grass

A number of species belonging to the genus having thin grass like stems that are erect to arching, carrying narrowly linear-lanceolate leaves are grown in gardens. *Arundinaria auricoma* has creamy yellow variegations, *Arundinaria fortunei* has a compact growth habit and leaves variegated with light green colour, *Arundinaria pumila* is a miniature grass with bushy habit and *Arundinaria suberecta* has thin stems attaining a maximum height of one meter that arch downwards gracefully, carrying thin and narrow light green leaves.

Arundo donax var. *variegata*

Variegated Giant Reed grass

Plants are tall, erect and have reed-like stems. The leaves are variegated.

Axonopus compressus
Syn. *Axonopus compressus var. australis / Milium compressum / Paspalum platycaule*

American Carpet grass, Blanket grass, Broad leaf Carpet grass, Louisiana grass, Tropical Carpet grass

Plants are hardy, thick and creeping-stemmed grasses having relatively broad leaves with slightly wavy margins. They are warm season low-maintenance grasses tolerant to shade and high water tables.

Axonopus fissifolius
Syn. *Axonopus affinis*

Caratao grass, Common carpet grass, Mat grass, Narrow leaved Carpet grass

Plants are hardy, relatively narrow leaved and thick stemmed. They provide a medium-coarse textured grass cover that needs close mowing.

Briza media

Quaking grass

Plants are tall and erect in habit. The flower heads are button-like and attractive.

Calamagrostis acutiflora

Feather Reed grass

Plants are tall and clump forming, attaining heights of up to 1.5m. They are tolerant to water logging and are ideal marginal plants of water gardens. The inflorescence is upright and attractively coloured red and white. In cv. 'Overdam', the foliage is edged with a white line and is spectacular.

Chasmanthum latifolium

Northern Sea Oats

Plants are shade loving perennials attaining heights of up to 1.2m.The leaves are upright and arching and become coppery brown in Autumn. They are also container-grown. Plants flower in summer and the seed bearing flower heads are coloured green and red and patterned like braids and have decorative value.

Coix lacryma-jobi

Job's tears

These grasses are grown as annuals. Their grains are globular, hard and pearly white. They are edible and are also used for making beads.

Cortaderia richardii

Black Pampas grass

Plants are tall, attaining heights of up to 2.5m. The leaves are green and the flower heads are black in colour, giving attractive contrasts when specimen planted. The shorter-statured 'Dwarf Pampas grass' has grayish green leaves and white flowers. The 'Pink Pampas grass' has pink flower heads. The 'Silver-stripe Pampas grass' has leaves that are variegated with silvery white lines and the 'Sun Stripe Pampas grass' has leaves with green margins variegated in yellow at the center.

Cynadon dactylon

Bahama grass, Bermuda grass, Devil's grass, Dog's tooth grass, Hariyali grass, Indian Dhoob grass

These are the most extensively grown among the lawn grasses of the tropics. They are high temperature tolerant and not cold hardy. A large number of varieties that differ in blade textures are grown. 'Sahara' and 'Yuma' are two cultivars that are mixed and planted. 'Princess-77' has darker green leaves and provides a thick grass cover with good texture and 'Yukon' has comparatively greater cold tolerance and a fine texture. Hybrid cultivars that are seedless and are propagated vegetatively include 'Tif Green', 'Tifton 419' and 'Tifway'.

Deschampsia cespitosa

Bronze Veil grass, Tufted Hair grass, Tussock grass

Plants are mound forming and fine textured. The leaves are dark green. The inflorescences are bronze-gold in colour and provide spectacular mass effect when clumped.

Eragrostis elliottii

Blue Love grass

Plants are bushy and produce creamy white cloud-like inflorescences above the level of foliage.

Erianthus ravennae

Hardy Pampas grass, Plume grass

Plants are upright, attaining heights of up to 3.5m and form clumps with a diameter of more than 1.5m. The inflorescences are feather like. The lower stems and leaves tend to break in high winds and hang down characteristically.

Erimochloa ophiuroides

Centipede grass

This is a grass indigenous to China and has a characteristic creeping habit. The stolons are short. The leaves are light green. Plants are adapted to acidic soils and yellowing due to iron deficiency is seen in non-acidic soils. They are recommended for establishing low-maintenance lawns in acidic soils and sandy areas.

Fargesia murielae
Syn. *Arundinaria murielae*

Umbrella bamboo, Bamboo grass

Plants are bushy perennial grasses attaining heights of up to 1.5 m or more. Their umbrella-like habit, strong stems and abundance of thick dark green foliage make them desirable plants for gardens.

Festuca ovina

Blue Fescue, Sheep Fescue, Fine Fescue

Plants are clump-forming, semi-evergreen perennial grasses that attain heights below 0.5m. The foliage is grayish – blue in colour. They are the most shade tolerant among the grasses. 'Large Blue Fescue' belongs to *Festuca amethystine.* The 'Tall Fescue'-*Festuca arundinacea* has wider leaf blades and rapid regrowth after mowing and is partially shade tolerant. The 'Silvery

Blue fescue' *Festuca glauca* is short, bushy and has thread like leaves. 'Creeping Red Fescue' *Festuca rubra* as well as the 'Hard Fescue' *Festuca longifolia* are used in mixed planting blends with 'Perennial Ryegrass' or 'Kentucky Blue grass'. *Festuca scoparia* the 'Bear-skin fescue' has dark green leaves and a fine hairy texture.

Glyceria maxima

Variegated Manna grass

These are hardy and adapted to wet land situations. The leaves have creamy white variegations and are tipped pink.

Hakonechloa aureola

Golden Ground-cover grass

Plants are short, bushy and decumbent. The foliage resembles animal hair.

Hakonechloa macra

Japanese Forest grass

Plants are adapted to moist partially shaded locations. They have a bushy habit.

Hierchloa odorata

Sweet grass

Plants are perennial aromatic grasses that are cold hardy and have a vigourous spreading habit. The leaves are braided and dried and burnt as incense.

Imperata cylindrica var. *koenigii*

Japanese Blood grass

The leaves are dark green and the upper ones turn dark red in summer. The flower heads are grayish white and provide a vivid contrast when mass planted.

Isolepis crenua

Fiber Optic grass

Plants are erect with thread like bushy foliage. The inflorescence stalks terminate in rounded white droplet-like clusters of flowers.

Koeleria glauca

Blue Hair grass

Plants have blue-gray leaves. The flower heads are golden yellow and are used as dried flowers.

Lagarus ovatus

Bunny Tail grass, Hares Tail grass, Rabbit Tail grass

Plants are compact in habit and produce a cluster of globose white tufted inflorescences resembling bunny's tails.

Leymus arinarius

Blue Lyme grass, Blue Dune grass, Sand Lyme grass

These are hardy, coarse textured grasses attaining heights of up to 1m. The foliage is bluish-green to grayish-green in colour. The stolons are invasive and require timely pruning to contain them.

Lolium perenne

Perennial Ryegrass, Winter Ryegrass

These are cold-hardy tufted long stemmed grasses that spread slowly by tillers or slips. Their pollen is a major cause of hay fever.

Melinis nerviglumis

Ornamental Savannah grass, Ruby grass

Plants are clump forming and attain heights of up to 60cm. They flower in late summer. The inflorescences carry arching heads of fluffy pink glumes.

Miscanthus sinensis

Japanese Silver grass, Porcupine grass

Plants attain heights of up to 1.5m and have a compact growth habit, forming dense clumps. The leaves are narrow, green and have a coarse texture. Cultivars having white or silvery white stripes and a reddish tinge on the leaves are grown. The inflorescences are white, pink or red tinged. Plants need to be cut back frequently to restrain growth.

Molinia littoralis

Purple Moor grass

These have a mound like habit and attain heights of up to 1.5m. The leaves are bright green. The inflorescences are held above the level of the foliage.

Muhlenbergia capillaris

Gulf Muhly, Hair-awn Muhly, Muhly grass, Purple Muhly

Plants are upright, attaining heights of up to 1.25m. The stems and leaves are thin and wiry giving an unusual soft and fine texture. This grass flowers in autumn and produces plumose panicles carrying reddish pink flowers. The seed bearing heads are purple in colour and provide a beautiful purple haze in gardens when mass planted.

Oplismenus hirtellus var. *vareigatus*

Miniature Variegated Ribbon Grass, Japanese Ribbon Grass

Plants have small ovate-lanceolate leaves with white variegations and long stoloniferous stems. They are suited for hanging pots.

Panicum virgatum

Lowland Switch grass, Tall Panic grass, Tall Prairie grass, Thatch grass, Wild Red-tip, Wobsqua grass.

These are perennial clump-forming stoloniferous grasses that attain heights of over one meter. The leaves are blue-green. The rust coloured inflorescences are attractive and eventually produce white fluffy seed heads.

Plants tend to be invasive and were previously the sources of the notorious 'switches' used as whips. Their tolerance to dry land conditions has targeted them as a potential bio-fuel and as plants suited for xeriscaping.

Paspalum notatum

Bahia grass, Dallis grass, Pensacola Bahia

These are low-growing grasses with creeping, stout stolons primarily grown for forage purposes. They are preferred as low-maintenance requiring turf grasses in particular areas.

Pennisetum alopecuroides

Perennial Fountain grass

These have a mound forming habit and carry white tipped flower heads that arch gracefully. They are used as 'Accent plants' in xeric landscapes. 'Purple fountain grass' or 'Annual fountain grass' *Pennisetum setaceum* has narrow red to purple tinged leaves and grows in clumps that need yearly replanting. *Pennisetum villosum* the 'Feather- tip Fountain grass' has fluffy white flower heads. 'Miniature fountain grass' attaining heights of up to 30cm is also grown.

Phalaris arundinacea

Ribbon grass

Plants attain heights of up to one meter and have gracefully arching ribbon like leaves. A number of cultivars having variegations are grown in gardens. Cultivar 'Picta' has green leaves with white stripes, 'Luteo picta' has leaves striped golden yellow and 'Feesey's Form' has a pink blush on the leaves.

Poa pratensis

Kentucky Blue grass, Smooth Meadow grass

The genus *Poa* has more than 20 species of grasses adapted to varying situations in the Northern temperate regions. Their seed heads are blue in colour. Kentucky Blue grass is sun loving and stoloniferous and performs well under good management. Seed germination requires two weeks. Excessive development of 'thatch' composed of accumulated old and dead stems and roots near the surface is a disadvantage of Kentucky Blue grass lawns, observed under high fertility conditions.

Sasa veitchii

Nagasa bamboo grass

Plants attain heights of up to 5m and can tolerate partially shaded situations. The leaves are ovate and have acuminate tips. The leaf margins become creamy white before winter. A dwarf cultivar 'Nana' is also grown.

Schizachyrium scoparium
Syn. *Andropogon scoparius*

Little Bluestem grass

Plants are upright arching perennials attaining heights of up to 1m. They have short spreading stolons and a deep root system. Plants flower in August-September and during this period, the leaves attain a bronze-red colouration.

Sporobolus heterolepsis

Prairie Droop seed

These are hardy, drought tolerant slow growing grasses having a dense mound of bushy leaves that are glossy green and thread or hair like. They become deep orange to coppery coloured in autumn. Scented panicles are produced in late summer and are held above the foliage.

Stenotaphrum secundatum

Charleston grass, St. Augustine grass

These are shade tolerant warm season grasses that are low growing. They produce strong stolons carrying flattened broad leaves with rounded tips. Variegated cultivars are ideal ground covers for brightening up shaded areas in gardens.

Zea mays

Striped corn, Variegated corn

These are variegated cultivars of edible corn that are grown for their ornamental and rarity value in gardens. Plants attain heights of up to 4.5m.

Cultivar 'Albovareigata' has creamy yellow and pink stripes, 'Harlequin' has deep red ears and red stripes and 'Red Stalker' has red and purple stripes.

Zoysia japonica

Japanese grass, Korean grass, Zoysia grass

These are one of the most widely used lawn grasses of the temperate regions. The leaves are dark green have a medium-coarse texture and plants are slow growing, cold hardy and thrive under good management. 'Meyer Zoysia' also called 'Super grass' is a medium-coarse textured selection having good spreading capacity as well as cold tolerance. Cultivar 'Emerald' is an interspecific hybrid of *Zoysia japonica* and *Zoysia matrella* and is fine textured and less cold hardy, having dark green leaves.

BAMBOOS

Bamboos are unique giants of the grass family that have been awarded a separate familial status - in *Bambusaceae* - by several systematists. Many bamboos indigenous to various geographical areas are integrated into the traditions and lifestyle of native populations. Many have been introduced into different parts of the world and are being evaluated for adaptability, longevity and economic potential. Many have been popularized as garden plants and a large number are grown for their spectacular display value. A 'Bambusetum' where collections of bamboos are maintained has become an integral part of many Botanic Gardens. They are also the most often used plants in the West, to symbolize the exotic East.

Ornamental bamboos grown as garden plants include the following:

Ampelocalamus scandens

Climbing bamboo

Plants are bushy with thin culms that tend to climb over neighbouring vegetation in native habitats in mountainous regions of China. The tips of culms tend to droop downwards.

Bambusa atra

Reed Bamboo

Plants are tall growing and have thin culms that appear reed like and grow in tight clusters. The persistent bracts covering them are attractive.

Bambusa bambos

Thorny bamboo, Illi Mula (Mal.)

These are tall growing thick culmed bamboos grown in India for multifarious uses. The culms are thorny.

Bambusa beecheyana

Beechy bamboo

Plants are tall growing with arching culms. The young shoots have a characteristic powdery white coating.

Bambusa chungii

Tropical Blue bamboo, Emperor's Blue bamboo

This is a famous Chinese bamboo that is tall growing and has pale blue culms. This colour is caused by the powdery bloom covering them.

Bambusa cornigera

Zig-zag bamboo

These are tall bamboos having culms that grow in a zigzag manner.

Bambusa dissemulator

Dragon's Nest bamboo

Plants grow in large dense clumps. They are mainly grown as a specimen plants in Bambusetums.

Bambusa emeiensis

Ornamental bamboo

Plants have smooth and straight culms that are entirely green or striated with yellow or dark green lines. They are tall growing and cold hardy.

Bambusa etuldoides

Asian Lemon bamboo

Plants are hardy and popularly grown in tropical gardens. The culms are yellow and have bold green stripes. The young culms have a pink colour.

Bambusa heterophylla

Variegated-leaved bamboo

Plants are short and bushy in stature. The leaves have light green variegations.

Bambusa intermedia

Variegated Black bamboo

Plants are tall growing and the culms are dark green with dark purple variegations.

Bambusa lako

Timor Black Bamboo

Plants are tall growing and arch gracefully at the tip. The green culms become shiny purplish-black on aging making these outstanding specimen plants.

Bambusa malingensis

Maling bamboo, Sea Breeze bamboo

These are dense clump forming bamboos with thick culms and small laves. They are salt tolerant and ideal for exposed landscapes.

Bambusa nana

Dwarf Feather-leaf bamboo

Plants have closely spaced culms that are branchless at the base. The leaves are thin and feathery.

Bambusa nutans

Nodding bamboo

Plants are short and bushy with very small leaves. The culms tend to nod when swayed by winds.

Bambusa oldhami

Oldham's bamboo

Plants are tall growing with straight and erect culms that are relatively soft and thick.

Bambusa polymorpha

Silk bamboo

Plants have dark green culms that are covered with dull white silken hair.

Bambusa sinospinossa

Chinese Thorny bamboo

Plants have small leaves and the culms are thin and have small basal thorns.

Bambusa textilis

Royal bamboo, Silk Bamboo

Plants are tall growing and vigorous and have bluish white bloom in patches on culms. A number of varieties with thin to very thick culms that stand more or less erect or arch gracefully are grown in the tropics.

Bambusa tulda

Punting Pole bamboo

Plants form large clumps. The culms are uniformly thick and grow straight up to considerable lengths and may be coloured dark green or have yellow striations.

Bambusa ventricosa

Buddha's Belly bamboo

Plants have dark green culms with short and characteristically swollen internodes. They are popular plants of tropical gardens. The cultivar 'Kimmei' has yellow culms with stripes.

Bambusa vulgaris

The Common bamboo, Ornamental bamboo

These are tall growing bamboos that form open clumps, grown throughout the tropics. Plants with long or discreet striations of dark green on yellow culms are also grown.

Chusquea liebmanii

Clumping Mountain bamboo

Plants grow in dense clumps and have thin arching stems that can clamber up mountainsides. They are cold hardy.

Dendrocalamus asper

Black Asper bamboo

Plants are large and have black culms that are used for timber.

Dendrocalamus giganteus

Giant bamboo

Considered to be one of the largest of bamboos, these are native to Burma. The culms attain unique proportions and plants are found as rare specimens in gardens of the tropics.

Dendrocalamus amoenus

Angel Mist bamboo

Plants grow in dense clumps and have a whitish bloom and irregular green stripes on culms. Plants are cold hardy.

Dendrocalamus sikkimensis

Sikkim bamboo

Plants form large clumps. The culms become rust red in colour with ageing.

Dendrocalamus strictus

Ever flowering bamboo, Male bamboo

Plants are tall growing, with slender culms and slightly drooping leaves.

Dinochloa malayana

Climbing bamboo, Vine bamboo

Plants are native to South East Asia. They tend to climb on to neighbouring vegetation attaching themselves with the help of culms that are rough on one side.

Drepanostachyum khasianum

Blue Mountain bamboo

Plants are relatively short in stature and grow in small clumps. The culms are dark green and have a powdery bloom that gives them a blue tinge.

Gigantochloa atroviolacea

Tropical Black bamboo

Plants are tall and have greenish black culms that open outwards like a fountain making them highly ornamental. A number species are grown as specimen plants. They include

Gigantochloa marga, the Tiger Stripe bamboo, which has unusual stripes, *Gigantochloa pseudoarundinacea,* which has strikingly variegated culms and *Gigantochloa ridleyi,* which has closely spaced culms with persistent sheaths.

Guadua angustifolia

Striped Guadua bamboo

Plants have large green culms with irregular yellow stripes.

Nastus latus

New Guinea bamboo

Plants have greenish yellow culms that deepen to yellow on ageing. The leaves are narrow and hang in a cascade making these very ornamental.

Otatea acuminata

Mexican Weeping bamboo

Plants are drought tolerant and moderately cold hardy. An abundance of linear lanceolate leaves cascade over the culms, providing an evergreen cover.

Schizostachyum brachycladum

Sacred Bali bamboo

Plants are adapted to the humid tropics. The culms are yellow in colour with discreet green striations. The leaves are large and a few have variegations.

Thyrostachys siamensis

Monastery bamboo

Plants are tall growing and have relatively slender culms that are almost solid. The leaves are small and ornamental.

SEDGES

Sedges belonging to the family **Cyperaceae** having ornamental value are grown in gardens especially in Water gardens and Marsh Gardens as marginal plants. Three-angled stems characterize them and they thrive in open, wet and humid habitats. Their bushy growth habit and dark green to bluish gray or coppery red foliage tints and the textural contrast they provide when compared to other non-graminaceous plants make them landscape plants of interest. They are also used in dry flower arrangements.

Some of the more commonly grown **ornamental sedges** include the following:

Carex buchananii

Leather Leaf sedge, Milk Chocolate sedge, New Zealand sedge

Plants are drought tolerant and attain heights of up to 60cm. The leaves are upright, light chocolate brown turning coppery brown in colour, with acuminate tips.

Carex comans

Bronze sedge, Small Red sedge

Plants are bushy and compact attaining heights of up to 30cm, with thin thread-like drooping leaves. The cultivar 'Frosted Curls' has arching leaves that are slightly curled.

Carex elata

Golden sedge

Plants attain heights of up to 70cm. The leaves are yellowish green, fine textured and arch gracefully. A number of cultivars with variegations of yellow and green are grown.

Carex flagellifera

Toffee sedge, Orange Hair sedge, Weeping Brown sedge

Plants have light to coffee brown leaves that arch gracefully and are ornamental.

Carex glauca

Steel-blue sedge

Plants grow upright and have erect bluish green leaves.

Carex grayi

Bur sedge, Gray's sedge, Mace sedge, Morning Star sedge

Plants are clump forming and have very showy star shaped seed heads.

Carex morrowii

Japanese sedge, Morrow's sedge

Plants are bushy, clump forming sedges adapted to marshy situations. The leaves are arching and become completely reflexed and have white margins. The cultivar 'Ice Dance' has bold white stripes and 'Silver Sceptre' and 'Silver Tassel' have silvery lines, 'Evergold', 'Goldband' and 'Variegata' have yellow variegations.

Carex nigra

Black Flowering sedge

The leaves are grayish green in colour. At flowering the flowers provide an ideal contrast with the leaves.

Carex pensylvanica

Pennsylvania sedge

Plants are small and delicate looking and are grown ornamentally.

Carex pilulifera

Pill sedge

Plants are dwarf in stature and are ideal ground covers for wet areas.

Carex tenuiculmis

Chocolate sedge

Plants have coppery brown leaves. Cultivar 'Cappuccino' is popularly grown.

Carex testacea

Orange sedge

Plants are bushy and have orange-red coloured leaves that are thread-like.

Carex trifolda

Sunrise sedge

Plants have attractive leaves with golden yellow leaf margins.

Cymophyllus fraserianus

Fraser's sedge

Plants are woodland sedges adapted to partial shade.

Cyperus alternifolius (see under Perennial Foliage Herbs)

Cyperus papyrus

Feather Duster plant, Papyrus reed, Papyrus sedge

Plants are native to the Nile delta. They are tall, leafless perennial herbs that produce clumps of angled shoots from thick woody stolons. These terminate in feather duster like shoots that eventually produce brownish green flowers and fruits. The reddish brown papery scaly bracts covering the bases of shoots are ornamental and plants are also grown for their historic value. *Cyperus nanus* is short in stature and is being popularized as a garden plant.

Eleocharis acicularis

Spike sedge, Needle sedge, Dwarf Hair grass

Plants are dwarf perennials attaining heights of up to 15cm. They are adapted to shallow watery habitats and produce very small white flowers. They are ideal for small water gardens.

Eriophorum lineatum

Cotton-ball sedge, Cotton grass

Fine bristles extending above the flowers that terminate in dense white ball-like tufts characterize plants, making them ornamentals of water gardens.

Uncinia egmontiana

Orange Hook sedge, Orange New Zealand sedge

Plants are erect to arching and form stiff and compact clumps. The leaves are dark orange coloured and deepen to a crimson-blond colour.

RUSHES

Rushes are native to shady woodlands and marshes of the temperate regions. They belong to the family **Juncaceae**. Most of them are shade adapted and unlike sedges, have cylindrical shoots that are solid.

Some of the more commonly grown **ornamental rushes** include the following:

Juncus effusus

Common rush, Soft rush

Plants have a dwarf spreading habit. The leaves are dark green and slightly twisted.

Juncus filiformis

Corkscrew rush

Plants are erect and have spiraling dark green leaves held above the water level.

Juncus glaucus

Blue rush

Plants are tall growing and have erect bluish green leaves.

Luzula nivea

Common Wood rush

The foliage is evergreen and finely hairy. The inflorescences appear in dense showy white clusters and droop attractively.

contd. colour plates

Bambusa etuldoides

Bambusa etuldoides 'Variegata'

Bambusa ventricosa

Fargesia murieliae

Nolina recurvata

Ophiopogon intermedius

Chapter 10

ORNAMENTAL WATER PLANTS

Ornamental plants that are adapted to an aquatic environment are commonly grown in water gardens and pools. They include those that are entirely submerged or floaters that are free living in the surface water or having roots anchored firmly in the ground beneath and marginal plants that thrive along the edges of water bodies and in shallow water. Spectacular flowering plants like the lotuses and water lilies to plants like the tape grass that are entirely submerged and grass like are grown in natural or structured fresh water habitats throughout the world. Many of these are also inhabitants of aquaria.

Some of the popularly grown **water garden plants** are listed below:

Acorus calamus Araceae

Sweet Flag, Vayambu

Plants are rhizomatous herbs adapted to low lying and wet situations. The leaves are closely spaced, distichously arranged and resembling a fan, leathery, dark green, smooth and shiny. The inflorescence is a spadix that emerges from the middle of the terminal leaf. Plants are valued medicinally for their aromatic rhizomes and are also commonly grown as marginals of water gardens.

Alisma gramineum Alismataceae
Syn. *Alisma geyeri / Alisma lanceolatum*

Narrow leaved Water Plantain, Ribbon leaved Water Plantain

Plants are corm-producing perennials adapted to fresh or brackish water. The submerged leaves are ribbon like and the surface leaves are stiff and ovate. Flowers are produced above or below the water level.

Alisma plantago-aquatica Alismataceae

Water Plantain

Plants are bushy, rhizomatous perennial herbs of marshy areas that are emergent with ovate leaves held on rounded petioles. The flowers have three white or pale pink petals.

The rhizomes are edible.

Alternanthera reinickii Amaranthaceae

Magenta Water hedge

Plants are submerged or emergent herbs with leaves that are elliptic lanceolate and colored red on the lower and green on the upper surfaces.

Ammania gracilis Lythraceae

Cognac plant, Red Ammania

Plants are adapted to acidic marshy areas. The shoots are thin and narrow and have wavy pale green to bronze red leaves attaining lengths of up to 20cm.

Anacharis canadensis Hydrocharitaceae
Syn. *Elodea canadensis*

American Water weed, Common Elodea

Plants are completely submerged aquatic herbs of fresh water habitats. They are dioecious. The leaves are sessile, whorled and crowded towards the tips of shoots. The leaf stalks detach from plants, float away and root enabling vegetative propagation. The flowers are small, white and open at the surface of their watery habitat.

Anubias barteri Araceae

Broad leaved Anubias

Plants are partially to fully submerged rhizomatous herbs having a compact growth habit. The leaves attain a length of up to 30cm.

Aponogeton crispus Aponogetonaceae

Ruffled, Crinkled Aponogeton

Plants are erect submerged rhizomatous herbs. The leaves are 20 to 30cm long and up to 6cm broad with wavy margins. The flowers are small, white or pink, scented and are produced on long stalks of up to 80cm length. *Aponogeton undulatus* has pale green leaves that are deeply undulating.

Azolla caroliniana Azollaceae

Caroline Mosquito fern, Water fern

Plants are fresh water aquatic ferns having small green fronds covered by fine hairs. Older fronds become reddish brown in colour. Plants are grown in water gardens for the velvety texture presented by the foliage cover and are also cultivated for their valuable nitrogen fixing capacity.

Bacopa caroliniana Scorphulariaceae
Syn. *Bacopa amplexicaulis / Obolaria caroliniana / Herpestes caroliniana*

Giant Red Bacopa, Lemon Bacopa, Water Hyssop

These have prennial creeping stems and succulent leaves that are lemon scented. The flowers are blue.

Bacopa monnieri Scorphulariaceae

Brahmi, Water Hyssop

Plants are creeping wetland herbs having small thick and ovate leaves. The flowers are small and white. Plants are valued for their medicinal uses.

Cabomba caroliniana Cabombaceae

Green Cabomba, Fanwort, Fish grass, Washington grass

Plants are submerged or floating in habit and inhabit fresh water. The submerged leaves are finely divided and arranged in pairs while the floating leaves are linear and alternate. Short fragile rhizomes detach easily, enabling quick spread of plants. Flowers are small and white, found on the water surface.

Crassula aquatica Crassulaceae

Common Pygmy weed, Water Pygmy weed

Plants exist as thin patches or mats in submerged to marginal aquatic habitats. Shoots are greenish yellow and turn reddish purple with age. The leaves are very small, fleshy and have pointed tips.

Crinum thaianum Amaryllidaceae

Water Onion

These are submerged amaryllids attaining heights of up to 1.5m having ribbon like leathery leaves. The flowers are similar to those of water lilies.

Cryptocoryne affinis Araceae

Crypts, Water Trumpet

These are native to tropical Asia. The leaves are lanceolate with serrate margins and are bluish green on the upper and violet coloured on the lower surface. The inflorescence stalk is erect and long. The spathe is reddish purple and coiled.

Damasonium alisma Alismataceae
Syn. *Damasonium stellatum / Damasonium polyspermum*

Star Fruit, Thrumwort

Plants are native to shallow aquatic habitats. The leaves are submerged or floating. The flowers are star like with three petals and have fringed edges and yellow bases.

Didiplis diandra Lythraceae
Syn. *Peplis diandra*

Water hedge, Water Purslane

These are aquatic annuals having linear lanceolate leaves held angularly in four rows. When grown in aquaria the leaf tips become reddish in strong light.

Echinodorus amazonicus Alismataceae
Syn. *Echinodorus bolivianus / Echinodorus brevipedicellatus*

Amazon Sword plant

Plants are erect and have sword like leaves that are submerged or emergent. Plantlets form on submerged nodes. Inflorescences emerge above the water surface, carrying up to nine white flowers.

Elatine ambigua Elatinaceae

Asian Waterwort

Plants have submerged creeping stems that root at nodes forming an intertwining carpet that turns reddish during the hot season. The leaves are paired and spoon like.

Eichhornia crassipes Pontederiaceae

Water Hyacinth

These are floating gregarious plants that have attained the status of weeds in many parts of the world. They are valued in water gardens for their showy lilac blooms.

Eriocaulon aquaticum Eriocaulaceae

Pipewort, Seven-angled Pipewort, White Buttons

Plants are semi aquatic perennials. The inflorescences are erect and have dense rounded clusters of small white flowers.

Eusteralis stellata Lamiaceae

Australian hygro, Star rotala

These are aquatics that attain a height of 50cm. The leaves are thin narrow and held like a brush.

Hydrilla verticillata Hydrocharitaceae

Hydrilla

Plants are submerged-stemmed leafy herbs common to the tropics and are popularly grown in aquaria and shallow ornamental pools..

Hydrocharis morsus-ranae Hydrocharitaceae

European Frog-bit

Plants are stoloniferous floaters adapted to shallow water. The leaves are heart shaped.

Hydrocotyle leucocephala Apiaceae

Brazilian Water Ivy, Pennywort

These are submerged or floating plants that have round and reniform leaves with broadly crenulate edges. Flowers have three white petals that are slightly cupped.

Hygrophila serpyllum Acanthaceae

Marsh Carpet

Plants are small low growing and adapted to open wet places. The leaves are small and the shoots spread slowly. The flowers are very small and white or pink in colour.

Lemna minor Lemnaceae

Duck weed, Lily pad

Plants have small pad like rounded leaves that float on the water surface. The roots too float freely.

Ludwigia repens Onagraceae

Creeping Ludwigia, Narrow leaved Ludwigia, Red leaved Ludwigia

Plants attain a height of up to 50cm in submerged aquatic habitats. The leaves are narrowly elliptic with acute tips. The young leaves are red in colour.

Myriophyllum elatinoides Haloragaceae

Water Milfoil, Christmas Tree plant

Plants appear brush like and have fine delicate leaves.

Najas graminea Najadaceae

Najas

Plants are fresh water annual herbs having dichotomously branching shoots.

Najas minor has smaller and more fragile shoots.

Nelumbo nucifera Nelumbonaceae
Syn. *Nelumbium indicum*

Indian Lotus, Sacred Lotus

Plants are popular throughout the world from very ancient times. The bluish green, emergent deltoid leaves are ornamental. The flowers are white, pink or yellow. The creeping rhizomatous stems, seeds and flower buds are edible.

Nymphaea spp. Nymphaeaceae

Water Lilies

Plants are perennial fresh water herbs with creeping rhizomes that are widely grown in the tropics. The rounded leaves are not emergent and have slightly wavy margins. They are night or day bloomers.

Nymphaea alba is the European White Water lily, *Nymphaea capensis* is the Cape Blue Water lily, *Nymphaea coerulea* (syn. *Nymphaea nouchali* var *coerulea*) is the Egyptian Blue Water lily, *Nymphaea lutea* is the Yellow Water lily, *Nymphaea odorata* is the white or pink flowered Fragrant Water lily, *Nymphaea rubra* is the Indian Red Water lily that has dark pink flowers and maroon-red leaves, *Nymphaea stellata* is the Indian Blue Water lily which is the national flower of Sri Lanka and is found throughout the warmer parts of India and *Nymphaea tetragona* is the small flowered Pygmy Water lily. Improved cultivars of different species grown include day-bloomers such as 'Blue Beauty', 'Peach Blow', 'St. Louis Gold' and 'Royal Purple' and night-bloomers such as 'Texas Shell Pink', 'Juno' and 'Red Flare'. Cultivars with flower colours that deepen or lighten with age include the yellow flowered 'Aurora' and the dark Pink 'Solfatare'. The cultivar 'Charles de Meurville' has red petals tipped with cream in the outer whorls. Cold hardy cultivars grown are 'Apple blossom Pink', 'Queen of Whites' and 'Sunrise'.

Nymphoides aurantiacum Menyanthaceae

Floating Water Lily

Plants are fresh water floating herbs. The flowers are small white or yellow.

Nymphoides cristata Menyanthaceae

Fairy Water Lily, Floating Heart

The leaves are heart shaped to ovate and have reddish brown mottling along the margins. The flowers are white and the petals have undulating margins.

Nymphoides indica Menyanthaceae

Water Snow Flake

The leaves are heart shaped and bright green in colour. The flowers are white and the margins of petals are deeply fringed.

Pistia stratiotes Araceae

Water Lettuce

Plants are stoloniferous floaters of fresh water habitats that attain the status of weeds in untended waterways. The sessile leaves with wavy margins are arranged in a close spiral forming an attractive rosette.

Ranunculus aquatilis Ranunculaceae

Water Crowfoot

Plants have small rounded lily-like leaves. They produce flowers in a profusion. The flowers have five white petals with yellow centers.

Rorippa nasturtium-aquatica Brassicaceae

Water Cress

Plants have shiny orbiculate leaves and are grown ornamentally and cultivated as a vegetable. However, they grow as weeds in neglected waterways.

Sagittaria subulata Alismataceae

Needle Sagittaria, Floating Arrowhead

Plants are low growing and adapted to acidic wetlands. They attain heights of less than 40cm and grow as dense rosettes with the help of runner shoots.

Trapa natans Trapaceae

Water Chestnut

Plants are floating herbs grown for their edible seeds and ornamental white flowers that open during the afternoon hours. The fruits are spiny.

Typha latifolia Typhaceae

Cat-tail Typha, Lesser Reed Mace

Plants are tall growing solid stemmed and adapted to marshy areas. They are an essential component of marsh gardens. The flower heads are used as dry flowers.

Vallisneria spiralis Hydrocharitaceae

Tape grass, Vallisneria

Plants are submerged stem less fresh water herbs that are popular oxygenators of aquaria and ornamental pools. The leaves are narrow and strap-like. *Vallisneria tortifolia* has markedly twisted and curled leaves.

Victoria amazonica Nymphaeaceae
Syn. *Victoria regia*

Amazon Lily, Giant Water Lily

These are native to the Amazon basin and have been introduced into gardens all over the world as unique giants of the Lily family. The very large and round leaves that float with upturned edges are beautiful. The leaves attain a diameter of more than 3m and are held on petioles of 7 to 8m length. The upper surface of the leaves is green and the lower surface is red and has prominent veins. The flowers attain a diameter of up to 40cm and are night bloomers. They are white at opening and are pink in colour the next night onwards. *Victoria cruziana*, the Santa Cruz lily has smaller leaves with flat margins and a finely hairy lamina with small outward projections.

contd. colour plates

Cyperus alternifolius

Damasonium alisma

Nelumbo nucifera 'Alba'

Nymphaea lutea

Nymphaea rubra

REFERENCES

1. Bole, P.V. and Vaghani, Y. 1989. *Field guide to the common trees of India*. Oxford University Press, Bombay, 125 p.

2. Brooks, J. and Beckett, K.A. 1987. *The Gardeners index of plants and flowers*. Dorling Kindersley Ltd., London, 272 p.

3. Davidson, W. 1988. *A Gardeners guide to flowering houseplants*. Salamander Books, London, 160 p.

4. Gilbert, R. 1988. *200 House plants anyone can grow*. Dorling Kindersley Ltd., London, 144 p.

5. Moore, D.M. 1991. *The Guinness guide to plants of the world*. Guinness Publishing Ltd., London, 256 p.

6. Pal, B.P. and Swarup, V. 1974. *Bougainvilleas*. Indian Council of Agricultural Research, New Delhi, 105 p.

7. Papworth, D. 1988. *A Gardeners guide to conifers*. Salamander Books Ltd., London, 160 p.

8. Randhawa, M.S. 1983. *Flowering trees*. National Book Trust, India New Delhi, 208 p.

9. Randhawa, G.S. and Mukhopadhyay, A. 1986. Floriculture in India. Allied Publishers Ltd., New Delhi, 656 p.

10. Saldanha, C.J. and Dhawan, J. 1984. Plants of India. Oxford and IBH Publishing Co. Ltd., New Delhi.

11. Santapau, H. and Henry, A.N. 1984. A Dictionary of the Flowering Plants in India. Council of Scientific and Industrial Research, New Delhi. 198 p.

12. http://www.us.ipni.org January to June 2008.

INDEX

A

B

C

D

E

F

G

H

I

J

K

L

M

N

Q

R

S

T

U

V

W

Y

Z

Kolibri 360 GmbH
Ferdinand-Jühlke-Straße 7
99095 Erfurt, Deutschland
produktsicherheit@kolibri360.de